THE ELEMENTS OF PROBABILITY
AND SAMPLING

The
Elements of Probability and Sampling

By

FRANK A. FRIDAY

(late Economic Adviser to Electric and Musical Industries, Ltd.)

BASIL BLACKWELL
OXFORD
1967

631 10920 X Cloth bound edition
631 10940 4 Paper bound edition

PRINTED IN GREAT BRITAIN IN THE CITY OF OXFORD
AT THE ALDEN PRESS
AND BOUND AT THE KEMP HALL BINDERY

CONTENTS

CONTENTS

TABLES

CHARTS

CHARTS

FOREWORD

FRANK FRIDAY and I were friends for very nearly twenty years. I am glad to provide this personal introduction to the book on which he worked, on and off, for something like ten years. He was tenacious in this purpose; his work for Electric and Musical Industries, Ltd., took a good deal of what the office clock would have marked as his 'own' time, and the book also had to survive increasing competition from various more public kinds of duty. I think it was on May 4th, 1962, that, telephoning me about something else, he said that he had at last worked through to the end of the book. He was then looking forward to a needed holiday and, after that, to getting everything ready for the publisher. *Dis aliter visum*. Friday died suddenly on the following Wednesday, May 9th, a few weeks before his fifty-second birthday.

His book has to go without the preface which he would have written had he lived. We may imagine that he would have said something about the importance and interest of the ideas of probability which underlie so many statistical procedures. No doubt he would also have discussed the usefulness which he hoped that his book would have for the kind of reader whom he had especially in mind—one whose interest in statistics, perhaps, might have come, like Friday's own, with the job in which he happened to find himself. Such a one may well wish to go beyond the compressed context of books serving 'part-time' students and more deeply into fundamental ideas lying behind the formulae he has to learn.

Although we lack Friday's own development of these themes, perhaps something which is relevant and which belongs to him may be caught in echo from what I shall say about him personally. Threads go out from this book which tie into many strands drawn together in Friday's life. Friday's character no less than his ability played a part in his becoming a statistician of standing, and his training in economic statistics was the foundation of his career as an industrial economist. Our own acquaintance began when he was with the Board of Trade as a war-time civil servant. Friday, involved in a survey which related to clothes rationing, had suggested to Mr. H. Leake, the Board's Chief Statistician, that they should call in an outside consultant. Mr. Leake turned to the Nuffield College Post-War Reconstruction Survey, where

I was then employed, and our Director, Mr. G. D. H. Cole, asked me to look into it. So I came to discuss the work with Friday. Without going into detail which I cannot remember, I can say that being *invited* to criticize I did so with gusto, as one might expect of a young academic. And I fear that I had little initial sympathy for the approximations which have to serve those who try to solve practical problems in practical time and with such data as are available.

I also remember that I learned a good deal about the actual subject-matter, and came to see something of the sensitivity which can guide the flair of a good statistician in this kind of work. As it happened, some of my criticisms and suggestions were found useful and Friday was appreciative. I was surprised when there followed an official letter of thanks to the Survey from Mr. Leake, which made it clear that Friday had gone out of his way to acknowledge the details of any help on my side of our transactions. Normal protocol did not call for more than general thanks. I found Friday's generosity the more remarkable because experience suggested that it was not abnormal for those with official responsibility to be apprehensive that such detailed recognition of outside help might reflect adversely on the skill and standing of those who had received it. I took advantage of later opportunities to see more of Friday himself.

With closer knowledge, this incident can be related to Friday's character and career. He felt that there was a general professional responsibility towards the difficult problems which can confront economists and statisticians. Answering any call freely himself, he had no touch of the *hubris* which might have prevented the help of others in his own work. This outlook is also to be found in the friends he made among those with whom he had professional contacts. There is a private discussion club which they founded and which grew out of informal contacts between a group of acquaintances who had become freely available to each other when any was concerned with something on which another could help. They then came to have regular monthly meetings where general discussion was sparked off by an opening paper by member or visitor. The club was meeting the day after Friday died and the other founder members then present caused a change of name to mark its connection with him. Friday would have prized this tribute from an association to which he was so greatly attached.

It would come easily to describe Friday as a self-made man but with his firm loyalties he was himself conscious rather of the advantages he had had, beginning, apart from the long encouragement and care his mother gave him, with Ealing Grammar School. (He died as he got back home from a meeting to plan its jubilee celebrations. He bequeathed the school its choice from the library which he enjoyed building up. His books marked the wider interests of the humane economist; the technical economics and statistics books, dictionaries and reference books were companioned by general literature and books concerned with the arts. Music was a regular pleasure and, after a fresh start in his mid-thirties, Friday became a remarkable and skilled painter.) There had been some struggle in his advancement in business and his professional training but, if one found him looking back, it was to remember some benefit got from working with someone, or some specific help or encouragement. One instance is especially relevant to the present narrative: I remember Friday speaking of his luck, as a very young office worker at E.M.I., in being under Brenda Stoessiger (now Mrs. R. A. Clapham), a one-time pupil of Pearson who both guided his first work on statistics and encouraged an interest in what the procedures meant. That was a brief experience but it launched him into the serious spare-time study which eventually brought him the London University Diploma in Economics and Social Science, after an examination in which he gained the Gilchrist Medal.

As already recorded, Friday's career at E.M.I. was interrupted by the war. He enjoyed his period at the Board of Trade and this led to approaches to stay in Whitehall or go elsewhere, but that loyalty which was to hold him to his old company until he died proved the stronger. On his return to E.M.I. he developed an economics and statistics department which was a pioneer in going beyond sales department problems to include matters affecting policy issues facing the company. While the set-up continued, this was a good training ground in industrial economics and, apart from those holding regular E.M.I. appointments, there are a number of economics graduates who think of Friday's old department with affection because of the welcome with which he took them, one or two at a time, for vacation experience of some aspects of business in practice, always with some piece of work of their own on a problem of interest to his company. In his final position as full-time Economic Adviser to E.M.I., Friday's long experience of this kind of work and his contacts outside the business stood him in good stead in the contribution which he made to internal and external policy.

Friday developed many academic connections and was regularly available for lectures or discussions arising from his experience and interests. I recall his visits to our Oxford graduate seminar in Economics of Industries, or its predecessor, which led directly to well-known papers by him on the pricing of television receivers and on economic forecasting problems. In fact he wrote more widely on the latter subject and he became a regular contributor to newspapers, etc., on economic questions generally. He enjoyed active participation in the Royal Statistical Society but he came gradually to agree with those who saw the need for a professional body with entrance qualifications determined by examinations to help directly in the training of the increasing numbers who were making a career in statistics and to set common standards. After attempts to get such functions developed within the older society, he played a large part in the foundation of the Institute of Statisticians, becoming member of its Council and chairman of its Education Committee. As *The Times* obituary noted: 'As a person who refused to compromise on any of his principles he helped to lay a sound foundation in the early formative years. . . .'

The Times obituary is captioned 'A Noted Industrial Economist'. His work in business, his professional contacts and the work for the Institute of Statisticians, his open relations with academic researchers and students, the articles which came out of this activity, made Friday well known and respected in the circles concerned and gave him peculiar personal status. During the last three years of his life, however, he became a public figure through his contribution to the defence of resale price maintenance. The two pamphlets which bear his name* are, apart from this book, his only non-article publications, and there were a large number of more transient contributions to the r.p.m. controversy. This involved a heavy diversion of his private time and energy, but any regrets which Friday had on that account were counterbalanced by his feeling a professional duty to do what he did no less than that which he felt towards the other work and writing which had to be set aside.

On this occasion I can only indicate the general circumstances of Friday's intervention in the debate. Until the burst of propaganda for a general ban on resale price maintenance, Friday's own attitude was simply that the practice might well be justifiable in particular industries, on the basis that, in the case of gramophone records where he had intimate knowledge, he saw good reasons for its being to the advantage of the public in general as well as to that of the manufacturers. Uncomfortable about sweeping propaganda assertions which began to appear in the newspapers in 1959–60, he was particularly moved to action by a serious pamphlet whose author gave, as 'probably rather cautious' conclusions, an estimate that the ending of r.p.m. would lead

---

* *Fair Trade* (*Resale Price Maintenance Re-examined*), by P. W. S. Andrews and Frank A. Friday, Macmillan, 1960; *Shops and Prices* (*Inquiry into Resale Price Maintenance*), by Frank A. Friday, Pitman, 1961.

to reductions in the prices of goods then affected by r.p.m., amounting to an average of about 5%, with a total national gain to consumers of an annual saving of some £180 million of their expenditure. This, as a 'cautious' *minimum* was a very striking, hard figure and its appearance had a big effect in newspaper and other reaction.

Friday carried out an independent inquiry which suggested that r.p.m. was probably not operative over so large an area as had been suggested. Further, the pamphlet had cited Canadian price reductions following a ban on r.p.m. as pointers to the expected reductions on which its overall estimates of the average price-fall in the United Kingdom were formulated. Friday went back to the original Canadian sources and found that the figures quoted for price reductions were not 'average reductions' in any ordinary sense; they were the average of the 'largest reductions from manufacturers' suggested retail prices' found in the inquiry, and with only a partial coverage. Then, since even a 5% average reduction of affected goods should show up in general retail results, Friday examined Canadian retail margins over a period after the ending of r.p.m., and showed there was 'no evidence . . . of anything like a drop of 5% of the retail price anywhere' nor of 'a general downward pressure on gross margins'.

From this brief example, it will be seen that Friday was not trying to 'prove a case from figures' but, rather, that he was in the professionally respectable position of showing that available statistics could be interpreted quite to the contrary of others' claims. I shall not go further now, except that I should perhaps note the trenchant criticism of statistical aspects of the Board of Trade departmental inquiry, whose report is still unpublished, which Friday made in his *Shops and Prices*. A reader of publications on the other side of the dispute will find shifts in argument and emphasis which seem to be due to Friday's work, unacknowledged. The author who put forward the 'cautious' minimum benefit of £180 million was later to say 'Resale price maintenance has little or no influence on total national money expenditure . . .' and 'It is, of course, impossible to measure the extent to which . . . retail prices would be altered with the abolition of r.p.m. in any particular trade, or *generally*'.*

Without going into other details of the controversy, the net effect was that Friday felt called to take a more active part where business men were finding a solid hostile front, whose statistical and theoretical backing seemed to him to have serious weaknesses. The more he looked at the

* B. S. Yamey, 'Resale Price Maintenance: Issues and Policies', *The Three Banks Review*, December 1960. (*My italics*.)

details of other industries besides his own, the more he came to believe that r.p.m. was of advantage in important cases and that it was fairly harmless otherwise. In the event, he was pleased to become honorary economic adviser to the Resale Price Maintenance Co-ordinating Committee which was guiding the defence of businesses interested in the practice. He thus got drawn into taking on a burden of meetings, etc., which must have been too heavy from the point of view of his health. The drive which urged him to this came from the passion of an earnest economist-statistician.

Ending this brief personal review of Friday's life, and turning back more directly to the book, I wish to thank those who have had a special part in its publication. Mr. Eric Shankleman, also an old friend of the author, read through the whole manuscript and gave advice on its presentation. Then we were fortunate that Mr. E. Brewer, formerly senior lecturer at the College of Technology, Oxford, and now with Messrs. Brewer, Sidebottom & Partners, agreed to act as editor. Besides his work of preparation, Mr. Brewer has also seen the book through the press. Miss Mary Lees, Friday's former secretary, gave indispensable help in the assembling of the manuscript and produced a complete typescript. We are indebted to Miss A. C. Black for the index. The author's family and friends would wish me also to mention the personal interest of Mr. Richard Blackwell, whom Friday had long intended to be his publisher.

Nuffield College   P. W. S. ANDREWS
Oxford
August 1966

EDITOR'S NOTE

The Editor's task has embraced the correction of the manuscript, the drawing of graphs and preparation for publication. Apart from the rearrangement of the material in the footnotes and the appendix starting on page 79, the work is as intended by Mr. Friday.

The work was done in the belief that the book will be of great value to students and to business men who wish to obtain a clear understanding of sampling and probability without recourse to the underlying and rigorous mathematical theory. Mr. Friday has adopted a step by step reasoning process that leads from the solution of problems of chance, that are known to have concerned gamblers and mathematicians in the fifteenth and sixteenth centuries, through combinations and permutations to the normal curve and confidence limits. This basic approach is then used to illuminate the meaning of probability and ramdomness and to show logically the inevitability and economy of sampling procedures.

The Editor would be grateful if readers would refer to him any errors or interesting omissions that might be taken care of if there is a further edition.

CHAPTER I

THE INSPIRATION OF GAMBLING

PROBABILITY implies lack of certainty and lack of certainty is attributed to chance. In the dictionary 'chance' is defined as unforeseen, accidental circumstances; it is popularly regarded as dependent on an unpredictable quality called luck. But to the statistician, chance is predictable; it conforms to laws; it has its own pattern. Chance is not chaos; it is order and, paradoxically, the greater the disorder the neater the order. The theory of probability provides a measure of how near certainty our information is and from this the theory of sampling has been developed. Today many important statistics are obtained from samples and not only are these results reliable, but they are often more trustworthy than could be obtained from a complete count.

Modern sampling technique was born in the gaming room, out of the marriage of mathematics with gambling problems. Gambling is as old as the hills. It is absorbing and provides, for a few, the opportunity of getting rich quickly. Those who gamble always consider they stand a reasonable chance of winning, otherwise they would not enter the game. The person who bets on red in roulette is sure that he has no less chance of winning than if he bet on black and very likely has persuaded himself that he has a better chance. There is much superstition in gambling. To the statistician, the probability of the ball falling in red can be given mathematically and the results over a long run can be forecast closely. To many gamblers the question whether the ball will roll into the red or the black compartments is decisively influenced by the decision of a black cat to cross their paths earlier in the day, black cats being specially omniscient and omnipotent compared with other cats. Broken mirrors, lucky and unlucky numbers, walking under ladders, the position of the stars at birth, birds evacuating their bowels over the human head—all this and more comprises the very special theory of probability believed by many people.

But gamblers sometimes pause and think. The division of stakes and deciding what are fair odds usually causes the halt. Unfortunately the absence of written records makes it difficult to be sure what was done by early mathematicians and gamblers on problems of chance. The different throws which can be made with three dice were discussed in a commentary (dated 1477) on Dante's *Divine Comedy* and another Italian, Luca Pacioli, attempted to solve for the first time in a book on mathematics the problem of the equitable division of stakes in an unfinished game. A mathematician named Cardan (1501–76), who was also an astrologer and a keen gambler, wrote a small book on games of chance which was published in 1663, long after his death. In prison for casting the horoscope of Jesus Christ, Cardan rashly forecast the date of his own death. On the appointed day he was in good health and had to commit suicide in order to make his prophecy come true.

In 1654 a French gambler of ability, Antoine Gornbaud, Chevalier de Mere, decided to take certain problems to his friend, Blaise Pascal (1623–62), a brilliant mathematician who soon afterwards at the age of 31 withdrew from ordinary life to become a recluse. Pascal discussed the problems with another mathematician, Pierre de Fermat (1601–65). The correspondence between Pascal and Fermat on the Chevalier's problems provides the foundation of the theory of probability.

One of the most important and certainly the most famous of these was the Problem of Points. Two players with equal chances of winning, each staking 32 pistoles (the French currency of that time), play a game of three points. How should the stakes be divided if the game is broken off before reaching the third point?

Pascal solved the problem by considering alternative possibilities:

Case no. 1. Suppose that one player (called A) has won 2 points and his opponent (called B) has 1 point. If A wins the next point he obtains all 64 pistoles; if B wins they each have equal points. Therefore A is certain of half the winnings and he may get all. B may get half the winnings or nothing. Pascal suggests that A should take his certain 32 pistoles plus half the remainder of the stake for which the chances are even. Thus, A should have 48 pistoles and B 16 pistoles if both players want to break off the game without playing for this last point.

Case no. 2. Suppose now that A has won 2 points and B has won none. If A wins the next point he gets his three points, and 64 pistoles; if B wins it, he is still a point behind A. Thus, at worst, A's position is that of case no. 1 when he is entitled to 48 pistoles and he stands equal opportunity of winning the remaining 16 pistoles. So A should be given 56 and B 8 pistoles if they want to break off the game without playing for any more points.

Case no. 3. We now assume that A has only one point and B none. Neither can win on the next throw. If A wins he possesses 2 points and, as argued in case no. 2, would be entitled to 56 pistoles. If, however, B wins they each have one point and their chances of winning the game become equal again and each would be entitled to claim 32 pistoles at that point. Therefore the division should be on the basis of giving A his certain 32 pistoles and half the difference between 32 and 56, i.e. A gets 44 pistoles and B gets 20 pistoles if they decide to break off the game when A has won 1 point and B none.

Fermat arrived at the solution to this problem by another route. He reasoned in this way: If A needs 2 points more to win and B requires 3 points, the game must be finished in at most, 4 more throws (5 throws altogether). The possibilities are that A will get 2 points or more and B 2 or less, in which case A would win; or A will get only 1 point or none, and B 3 or 4 points, in which case B will win; Fermat listed the alternative answers:

Alternatives	Throw 1	Throw 2	Throw 3	Throw 4	Winner
(1)	A	A	A	A	A
(2)	A	A	A	B	A
(3)	A	A	B	A	A
(4)	A	A	B	B	A
(5)	A	B	A	A	A
(6)	A	B	A	B	A
(7)	A	B	B	A	A
(8)	A	B	B	B	B
(9)	B	A	A	A	A
(10)	B	A	A	B	A
(11)	B	A	B	A	A
(12)	B	A	B	B	B
(13)	B	B	A	A	A
(14)	B	B	A	B	B
(15)	B	B	B	A	B
(16)	B	B	B	B	B

Every one of the alternatives which show A as winning on two or more throws mean that A wins the game. In alternatives (1), (2), (3) and (4) he would win his third point on the second throw (remember that he has already won a point; these are the remaining throws). In alternatives (5), (6), (9) and (10) he would win the game on the third throw; in (7), (11) and (13) he would win on the last throw. B would win in alternatives (8), (12), (14), (15) and (16). Thus 11 cases are favourable to A and 5 to B. If, therefore, A has one point and B none and the players wish to divide the stake without playing on, the total stake of 64 pistoles should be divided in the ratio of 11 to 5 i.e. $\frac{11}{16}$ to A (44 pistoles) and $\frac{5}{16}$ to B (20 pistoles). This agrees with Pascal's answer.

The solution to the problem of points is essentially an analysis of all the alternative combinations of *equally likely results*. Pascal carried this much further in a treatise on the arithmetical triangle which was printed in 1654 although not published until 1665, three years after his death. The Pascal triangle was not new—from early times men had been fascinated by these numbers—but its application to the problem of points and to the development of the theory of combinations was certainly new.

A specimen of this triangle is shown in Table 1. The numbers are formed very simply. Beginning with units we get a series of natural numbers, thus:

$$
\begin{array}{ll}
1 & 1 = 1 \\
1 & 2 = 1+1 \\
1 & 3 = 1+1+1 \\
1 & 4 = 1+1+1+1 \\
1 & 5 = 1+1+1+1+1 \\
1 & 6 = 1+1+1+1+1+1
\end{array}
$$

Next, the natural numbers are added downwards giving at each stage another series known as triangular numbers:

$$
\begin{array}{ll}
1 & 1 = 1 \\
2 & 3 = 1+2 \\
3 & 6 = 1+2+3 \\
4 & 10 = 1+2+3+4 \\
5 & 15 = 1+2+3+4+5 \\
6 & 21 = 1+2+3+4+5+6
\end{array}
$$

Adding the triangular numbers together we obtain a series of new triangular numbers:

$$
\begin{array}{ll}
1 & 1 = 1 \\
3 & 4 = 1+3 \\
6 & 10 = 1+3+6 \\
10 & 20 = 1+3+6+10 \\
15 & 35 = 1+3+6+10+15 \\
21 & 56 = 1+3+6+10+15+21
\end{array}
$$

And so on to numberless orders of these series. As we shall see later this triangle has interesting properties. With it Pascal found the general answer to the Chevalier's problem of points, whatever the number of points already gained by each player. In this general solution we must use symbols. If A requires *a* points in order to win the game and B requires *b* points, the required ratios will be the sum of the first *b* terms for A to the sum of the remaining *a* terms for B in the row containing *m* terms, where $m = a+b$. Applying this to the example taken by Pascal and Fermat of an unfinished game for three points where A needed 2 points and B 3 points to win, the row containing 5 terms must be referred to. The first 3 terms $1+4+6$, total 11 and the remaining 2 terms, $4+1$, total 5, so the division of the total stake would be $\frac{11}{16}$ to A and $\frac{5}{16}$ to B. This agrees with the previous answer.

At first sight this may smack of magic, like those puzzles which tell us our ages by the addition and subtraction of what appear irrelevant figures. But, as with those puzzles, there is an explanation. Pascal was able to demonstrate that these series provided a measure of the different ways in which combinations could be made. This will be the subject of the next chapter.

CHAPTER II

COMBINATIONS AND PERMUTATIONS

WE have seen that the measurement of a player's chance of winning requires finding the total number of alternative possibilities having equal chances of occurring. The assumption of equality is essential because the division of the stakes is expressed as a fraction representing the chance of winning (e.g. $\frac{11}{16}$) and this would be meaningless if the units of measurement were not identical.

If a coin is tossed in the air it can land with one out of two sides uppermost. So far as one can see, there is no reason why the coin should fall on one side more than another. Therefore, *a priori*, we assume the chances to be equal and because there are only two possibilities, the chance of obtaining head is taken as one-half of the possibilities.

If two coins are tossed there are four possibilities:

First coin	Second coin
Head	Head
Head	Tail
Tail	Head
Tail	Tail

Thus we should say that the probability of getting two heads when throwing two coins would be one-quarter because two heads represent one possibility out of four equal possibilities.

Taking three coins we find 8 possible answers:

First coin	Second coin	Third coin
Head	Head	Head
Head	Head	Tail
Head	Tail	Tail
Head	Tail	Head
Tail	Head	Head
Tail	Head	Tail
Tail	Tail	Head
Tail	Tail	Tail

The probability of obtaining three heads in a throw of three coins is taken to be one-eighth because there is only one possibility of getting three heads out of eight alternative combinations.

6

The results of tossing four and more coins can be worked out in a similar way and the probability of obtaining a particular combination found by expressing the numbers of results containing the required combinations as a proportion of the total number of possibilities.

It is a convention of probability theory that probability is expressed in the form of a proportion. Thus a probability equal to nought means that there is no chance whatever of getting the stated result—for example, the probability = 0 that 4 heads will be obtained from tossing 3 coins; and a probability equal to unity means that all possible answers contain the desired result—for example, the probability = 1 that at least one head or one tail will appear when three coins are tossed because every possible combination contains one or the other attribute, head or tail. A probability equal to 0 or 1 is really certainty. We know for certain that the next throw of three coins can never be 4 heads and we know that there will always be a head or a tail uppermost in every throw. We have no doubt about it whatever. But we do not know whether the next throw will produce a combination of 2 heads and 1 tail. All we know is that it can happen because 3 such combinations are possible out of 8 equally likely combinations. Probability is mathematically defined as that proportion (expressed as a decimal) of all the possible equally likely results which have certain required properties or characteristics. Thus, the probability of getting 2 heads and 1 tail is 0·375 (or $3 \times \frac{1}{8} = \frac{3}{8}$).

This is all very well, but if we had 100 coins and tossed them how should we arrive at the probability of obtaining 95 heads and 5 tails? Listing all the possible alternative combinations would be a laborious task. We want a quicker route to the same answer.

The first step is to state the possibilities in the form of what statisticians call a *frequency distribution*. To show every individual item separately is unnecessary and when there are a large number of such items the result is thoroughly confusing. In a frequency table the items are grouped according to characteristics which they have in common: aggregates are obtained; and we get a series of totals instead of a mass of individual items. The results of the throws with two and three coins shown on page 6 can be classified in this way:

Results of throwing 2 coins	Frequency	Probability
2 heads	1	0·25
1 head and 1 tail	2	0·50
2 tails	1	0·25
Total combinations	4	1·00

Results of throwing 3 coins	Frequency	Probability
3 heads	1	0·125
2 heads and 1 tail	3	0·375
1 head and 2 tails	3	0·375
3 tails	1	0·125
Total combinations	8	1·000

These frequencies, it will be noted, are the same as the horizontal series in the arithmetical triangle. The Chevalier's problem dealt with consecutive throws and the above figures relate to a number of coins in one throw, but this makes no difference. One throw of three coins is just the same as three throws with one coin. The fact that three coins are used and are thrown simultaneously does not affect the result at all, provided the probability of getting heads is the same on all coins and the results are not related to one another in any way. These provisos clearly apply here. If the coins are not biased, there is no reason for assuming that the probability of getting heads is not 0·5 for all three coins and the results of throwing one coin do not affect the results of throwing the others. Similarly there is no reason to assume that one coin changes its probability on each throw, or having been heads on the preceding throw, the coin remembers this and tries, as some people seem to assume, to be tails next time. Therefore we can restate the above tossings in terms of successive throws of one coin:

Results of 4 throws	Frequency	Probability
4 heads	1	0·0625
3 heads and 1 tail	4	0·2500
2 heads and 2 tails	6	0·3750
1 head and 3 tails	4	0·2500
4 tails	1	0·0625
Total combinations	16	1·0000

If, in the problem of points, we say that A required two more heads in order to win and B required three tails, the probabilities of each winning are 0·6875 for A and 0·3125 for B, i.e. the 11 : 5 ratio discussed in the previous section. But now we know, not only how the probabilities are arrived at, but the meaning of each of the series in the Pascal triangle. The row containing the three terms, 1, 2, 1, represents the different ways of combining two objects or events, taking them in different ways; the row containing four terms, 1, 3, 3, 1, represents the different ways of combining three objects or events; the row containing five terms, the combinations of four things, and so on.

Let us examine these series closely, taking the combination of 4 coins, to begin with. In how many ways can 4 heads be combined 4 at a time? The answer is clearly once only. Next, in how many ways will three heads appear with 1 tail when a coin is thrown 4 times, or when 4 coins are thrown once? We know the answer to be four.

Now this is the same thing as saying that 4 different objects (say 4 different coins—a florin, a shilling, a sixpence and a penny) can be combined 3 at a time in 4 different ways. Assume that the 4 objects are A, B, C and D. Taking them in groups of 3 we get the following:

$$
\begin{array}{ccc} A & B & C \\ A & C & D \end{array} \qquad \begin{array}{ccc} A & B & D \\ B & C & D \end{array}
$$

Again look at it from the point of view of the *order* of results. A is a success (e.g. obtaining heads) at the first throw, B on the second throw, C, the third, D, the fourth. The four ways of combining 4 things three at a time is therefore exactly the same as the different ways of having 3 successes and 1 failure in the problem of points.

Taking the 4 in groups of 2 we get 6 combinations:

$$
\begin{array}{cc} A & B \\ A & C \\ A & D \end{array} \qquad \begin{array}{cc} B & C \\ B & D \\ C & D \end{array}
$$

The symbol which is used in the theory of probability to indicate combinations is simply a capital C, and $^{n}C_{r}$ means the combination of n objects taken r at a time. The row containing 5 terms in the triangle can therefore be written symbolically as a series of combinations:

$$
\begin{array}{ccccc} ^{4}C_{0} & ^{4}C_{1} & ^{4}C_{2} & ^{4}C_{3} & ^{4}C_{4} \\ 1 & 4 & 6 & 4 & 1 \end{array}
$$

This means that there is only one way in which there can be no successes out of four throws (i.e. 4 tails and no heads); there are 4 ways of getting only 1 success in 4 throws (clearly 1 success could be only first or second or third or fourth); there are 6 ways of obtaining 2 successes in 4 throws; 4 ways in which 3 successes at a time can be obtained in 4 throws; and one way of getting all successes (i.e. 4 heads in succession).

The arithmetical triangle can therefore be written in symbols which indicate the combination represented by each figure:

m	Combinations
1	$^0C_0$
2	$^1C_0$ $^1C_1$
3	$^2C_0$ $^2C_1$ $^2C_2$
4	$^3C_0$ $^3C_1$ $^3C_2$ $^3C_3$
5	$^4C_0$ $^4C_1$ $^4C_2$ $^4C_3$ $^4C_4$
6	$^5C_0$ $^5C_1$ $^5C_2$ $^5C_3$ $^5C_4$ $^5C_5$

This tells us what combination each figure in the triangle means but there is a limit to the value of the triangle for ascertaining combinations. In Table 1 the base row shows that there are 16,777,216 different ways of combining 24 things (objects or events) and that taking them 8 at a time can be done in 735,471 different combinations. This can be found quickly, but it is still very laborious to find out the probability of getting 95 heads and 5 tails in one toss of 100 coins, or 100 tosses of one coin, because we need to know *first*, the sum of all the ways of combining 100 things and *secondly*, how many different ways 95 successes can be obtained in 100 throws ($^{100}C_{95}$). Table 1 could be expanded of course, but the figures become so numerous and individually so large that an enormous area of paper would be required to give the complete triangle. We must arrive at combinations some other way. The value of the triangle is that it provides the clue to the formulae for obtaining combinations. What we need is a formula for arriving at ^nC_r without having to do all the additions of the arithmetical triangle. Pascal found that the triangle gave the clue to this formula.

There is really nothing frightening about a mathematical formula and the formulae which follow will be readily understood if the reasoning leading to them is studied carefully. A formula is merely a general rule. It is an important function of mathematics to provide solutions which have a general application, so that if we know the value of x then the value of another symbol, e.g. y, in the statement can be ascertained.

Let us refer again to page 4 where the formation of each series of numbers in the triangle was described. The first series of triangular numbers was formed from the natural numbers, thus:

$a =$	1	2	3	4	5	6	7
$T' =$	1	3	6	10	15	21	28
$^nC_r =$	$^2C_2$	$^3C_2$	$^4C_2$	$^5C_2$	$^6C_2$	$^7C_2$	$^8C_2$

We want a general rule which tells us that $T' = ka$, where a is the natural number and T' the triangular number we want to find. k is just a symbol representing the fact that some general rule, some equation, will link the two together.

It will be seen that each triangular number can be calculated by taking its corresponding natural number and multiplying it by half the next higher natural number. If the natural number is 3, the triangular number corresponding to it will be $3 \times \frac{4}{2} = 6$. If the natural number is 6, the corresponding triangular number is $6 \times \frac{7}{2} = 21$. Using symbols we can express this in a general formula:

$$T' = a \times \frac{a+1}{2}$$

$$= T' = \frac{a \times (a+1)}{2}.$$

This gives us a general rule for arriving at the combinations of any number of things taken 2 at a time ($^nC_2$). The number of things, it will be noted, equals one more than the natural number, i.e. $n = a+1$. The above formula can therefore be rewritten:

$$^nC_2 = \frac{n(n-1)}{2}.$$

Thus:
$$^8C_2 = \frac{8 \times 7}{2} = 28.$$

The next step is to calculate a general formula for arriving at the next set of triangular numbers:

a	1	2	3	4	5	6	7
T'	1	3	6	10	15	21	28
T''	1	4	10	20	35	56	84
$^nC_3$	$^3C_3$	$^4C_3$	$^5C_3$	$^6C_3$	$^7C_3$	$^8C_3$	$^9C_3$

The connection between second-order triangular numbers and the natural numbers can be similarly discovered. This time we find that each second-order triangular number can be obtained by taking its corresponding first-order triangular number and multiplying it by one-third of the next but one higher *natural* number, If $T' = 6$, then $T'' = 6 \times \frac{5}{3} = 10$; if $T' = 21$ then $T'' = 21 \times \frac{8}{3} = 56$. The general rule is thus:

$$T'' = T' \times \frac{a+2}{3}.$$

However we know that $T' = \dfrac{a \times (a+1)}{2}$ therefore

$$T'' = \frac{a \times (a+1)}{2} \times \frac{(a+2)}{3}$$

$$= \frac{a(a+1)(a+2)}{2 \times 3}.$$

This rule relates to combinations taken 3 at a time. The n in $^nC_3$ is always equal to 2 more than the natural number a so the formula can be rewritten:

$$^nC_3 = \frac{n(n-1)(n-2)}{3 \times 2}.$$

The third order of triangular numbers will be found to follow the same pattern:

a	1	2	3	4	5	6	7
T'	1	3	6	10	15	21	28
T''	1	4	10	20	35	56	84
T'''	1	5	15	35	70	126	210
$^nC_4$	$^4C_4$	$^5C_4$	$^6C_4$	$^7C_4$	$^8C_4$	$^9C_4$	$^{10}C_4$

In order to obtain T''' we take T'' and multiply by one-quarter of the corresponding number plus 3. Combining this with the formula for T'' the general rule for T''' is easily shown to be:

$$T''' = \frac{a(a+1)(a+2)}{2 \times 3} \times \frac{(a+3)}{4}$$

$$= \frac{a(a+1)(a+2)(a+3)}{4 \times 3 \times 2}.$$

The third-order triangular numbers refer to combinations of any number of things taken 4 at a time. Examining the natural numbers we note that n is always equal to its corresponding number plus 3, i.e. $n = a + 3$. Substituting $^nC_4$ for T''' the general rule becomes

$$^nC_4 = \frac{n(n-1)(n-2)(n-3)}{4 \times 3 \times 2}.$$

Our rule for getting the number of combinations is not, however, general enough yet. We have found how to arrive at $^nC_2$, $^nC_3$, and $^nC_4$ but we want to make this apply to any combinations.

Very little study is necessary to see how this can be found. In each of the above formulae the *numerator* consists of a figure which is equal to the total number of items being combined (i.e. n) multiplied by the preceding numbers ($n-1$, $n-2$, $n-3$, and so on) until there are as many numbers being multiplied as there are items being combined at a time. As the number of things taken at a time is symbolized by r this means multiplying n by all the preceding numbers down to $n-r+1$. For example, if $r = 5$ and $n = 100$, the numerator will be $100 \times 99 \times 98 \times 97 \times 96$. The denominator is simply r multiplied by all the preceding numbers back to 1, thus: $r(r-1)(r-2) \ldots 2 \times 1$. Whether we include or exclude the 1 makes no difference. The general formula we have been seeking showing how to calculate the number of combinations of n things taken in groups of r is therefore:

$$^nC_r = \frac{n(n-1)(n-2)(n-3) \ldots (n-r+1)}{r(r-1)(r-2) \ldots 2 \times 1}.$$

A number multiplied by all its preceding numbers right back to 1 is known as a *factorial* number and is written with an exclamation mark after it thus: $6! = 6 \times 5 \times 4 \times 3 \times 2 \times 1$. So the denominator in the above formula could be written more simply as $r!$ It would be spoken of as factorial r.

Factorial numbers are interesting. They were known 300 years before the birth of Christ but their meaning as measures of the number of different ways in which things can be arranged was not realized until the time of Fermat and, in particular, of James Bernoulli (1654–1705) who gave such arrangements the name by which they are known today —permutations.

There is a distinct difference between permutations and combinations. A combination takes cognizance only of the kinds of things included whatever their order. ABC and CBA are the same combination. The number of items must be taken into account so that BBAC although the same combination as ACBB is certainly not the same combination as ABC. A permutation takes into account every different *order* in which the *same* things—i.e. *one* combination—can be arranged. Thus, 3 letters of the alphabet can be combined 3 at a time only once ($^3C_3 = 1$) but they can be arranged in 6 different orders: ABC, ACB, BAC, BCA, CAB, CBA. Each is a different permutation.

The reason why factorial numbers give the number of permutations of a number of objects will easily be seen. Let us keep to the illustration of 3 letters of the alphabet, A, B and C. We can arrange these so that any one of the 3 is first but once a letter has been chosen for the first place, only two remain for the second place. Thus for *each* first letter there can only be two second places; if A has been selected first then only B or C can be second; and if B is taken as the second letter only C remains as the third. Thus, for each of the 3 first places there are 2 second choices—i.e. 3×2—and for each first and second places (i.e. 6) there can be one left for third place leaving the number of permutations of 3 letters in groups of 3 as $6 \times 1 = 6$.

The symbol used to denote permutations is quite simply the capital letter P, and as with combinations, ^nP_r means the permutation of n things taken r at a time. A factorial number $n!$ therefore means ^nP_n.

What we have next to decide is how to get ^nP_r. Again the answer can be readily demonstrated in the same way. If we want to know the number of permutations of 5 letters of the alphabet (A, B, C, D and E) taken in groups of 3, we can select any one of the 5 for first place. Once the letter has been selected we can only fill the second place from the remaining 4 letters and once we have our first and second letters we can take only one out of the remaining 3 letters to complete the groups of 3. Only permutations of 3 letters are required so we do not go right back to one, otherwise we should be taking the 5 letters 5 at a time ($^5P_5$). Therefore we stop at the third place and $^5P_3$ is $5 \times 4 \times 3 = 60$.

The procedure is shown in Chart 1. It will be observed that having selected A for the first place only B, C, D or E can fill the second place. If we have taken A for first and B for the second letter only C, D or E can be third. Similarly, if B has been selected as the first letter in the group, only A, C, D or E can be second and if D is made to fill that place only A, C or E can be third. Thus, 5 letters can be placed first and for *each* the remaining 4 letters can be placed second, giving $5 \times 4 = 20$ permutations. This represents $^5P_2$. Now for *each* of these permutations, for each decision as to the first and second places, there are 3 alternatives for third place, making $20 \times 3 = 60$ permutations ($5 \times 4 \times 3$). This is $^5P_3$, and we could go on in the same way and find $^5P_4$.

The general rule for calculating ^nP_r will be seen immediately. If n is the number of objects and r the number taken at a time, we have to multiply n by all preceding numbers down to the term before $n-r$, i.e. $n-r+1$. This is really the same as factorial n (n !) divided by the tail end which we do not want, viz. $(n-r)$!

$$^nP_r = \frac{n!}{(n-r)!} = n(n-1)(n-2)\ldots(n-r+1).$$

The number of permutations of n things taken r at a time is always larger than the number of combinations and the reader will see that there is a connection between the two formulae. Chart 1 shows this connection clearly. If we count the permutations of the letters A, B and C we note that there are 6; similarly with the 3 letters A, E, D. This is precisely what we should expect. It is the permutation of 3 letters taken 3 at a time ($^3P_3$) which is 3 ! = 6. All told there are 10 combinations of 3 letters out of the 5 ($^5C_3$) and for each there are 6 permutations. Therefore $^5P_3$ is really $^5C_3$ multiplied by $^3P_3$.*

* If the reader will refer to the combination formula on page 13 he will see that we get:

$$^nP_r = {}^nC_r \times {}^rP_r = \frac{n(n-1)(n-2)\ldots(n-r+1)}{r!} \times r!$$

$$= n(n-1)(n-2)\ldots(n-r+1)$$

which is the same as we previously obtained. Conversely, this means that a combination of n things taken r at a time is simply the total number of permutations of n things in groups of r with all the permutations of each combination of r letters eliminated:

$$^nC_r = \frac{^nP_r}{^rP_r} = \frac{n(n-1)(n-2)\ldots(n-r+1)}{r!}.$$

As ^nP_r can also be written as $\dfrac{n!}{(n-r)!}$ the formula for ^nC_r can be written in the

following form $\dfrac{n!}{(n-r)!} \div r! = \dfrac{n!}{(n-r)!} \times \dfrac{1}{r!} = \dfrac{n!}{(n-r)!\,r!}$.

C

CHAPTER III

IN WHICH THE READER IS APPEASED

BY now the reader will be really exasperated. First, he is informed that listing all the possible alternative results of tossing many coins would be so laborious that a quicker method is needed. It is next demonstrated that the arithmetical triangle consists of series of combinations, which is why Pascal was able to use it in the solution of the Chevalier's problem of points. But this cannot be used for large numbers, which are the really difficult cases, so a general formula for calculating combinations direct is required. And what do we get? . . . a frightening new number called a factorial which means that, in order to ascertain the number of ways in which 95 heads and 5 tails would result from tossing 100 coins all we have to do is make the following calculation:

$$^{100}C_{95} = \frac{100 \times 99 \times 98 \times \ldots \times 6}{95 \times 94 \times 93 \times \ldots \times 3 \times 2 \times 1}.$$

The probability of getting 95 heads and 5 tails would be merely this answer expressed as a proportion of the sum of $^{100}C_0 + {}^{100}C_1 + {}^{100}C_2 + {}^{100}C_3 + \ldots + {}^{100}C_{98} + {}^{100}C_{99} + {}^{100}C_{100}$. So a calculation containing a factorial would have to be made, presumably, for each combination!

Moreover, the reader must feel somewhat confused about combinations and permutations. On pages 9 and 10 the results of coin tossing are described as combinations, but there is surely a sense in which they are permutations. Are we not arranging two things, heads or tails (or success or failure) in groups according to the number of coins or throws? And are not arrangements called permutations and not combinations?

All this is true. We have certainly not yet obtained a quick method of calculating probabilities. For the most part the mathematicians dealing with these problems were amusing themselves. Very large numbers did not always enter the problems but when they did these mathematicians were anxious to find short cuts to the answer. They had no calculating machines. Such things are invented only when there is some need for them. Much of our knowledge begins with idle curiosity. The consequences may be dramatic and far-reaching. The necessity for developing

16

easy means of applying the knowledge becomes, as the proverb has it, the mother of invention.

Factorial numbers certainly take time to calculate. Their size, too, becomes unmanageable. For example $30! = 2 \cdot 6525286 \times 10^{32}$ which means that there will be 33 significant figures in the answer and the factorial of 30 is

$$265,252,860,000,000,000,000,000,000,000,000.$$

Factorials once calculated can be more readily employed by converting them into logarithms. At the time when Pascal and Fermat were playing with the gambling problems of the Chevalier de Mere, logarithms had only recently (in 1614) been invented by the Englishman, John Napier. But logarithms, like factorials, take a lot of calculating and have to be approximations. Thus the four-figure logarithm of $30!$ is $32 \cdot 4237$ and the antilogarithm will therefore be a number of 33 figures beginning with 2653, i.e. the same as the factorial given above but more approximate. They are, however, much more easily handled.

All this assists the calculation of combinations, which can be derived entirely from factorials by using the formula, $^{n}C_{r} = \dfrac{r!}{(n-r)\,r!}$. The number of possible combinations of 95 heads and 5 tails in the tossing of 100 coins is:

$$^{100}C_{95} = \frac{100!}{5!\ 95!}.$$

Taking the logarithms of these factorials we get the following simple calculation:

Log 100 !		157·9700
Deduct:		
Log 5 !		2·0792
Log 95 !	148·0141	
Log 5 !+log 95 !		150·0933
Log $^{100}C_{95}$		7·8767

Anti-log of $7 \cdot 8767 = 75,280,000$.

Thus the number of combinations each consisting of 95 heads and 5 tails is as many as 75,280,000. This still does not tell us the probability of getting this result. To get that we need also to know the total of all the possible results—$^{100}C_{0} + \ldots + {}^{100}C_{100}$.

Now it is not, in fact, necessary to sum *all* the alternative combinations of n things in order to find the total number of equally likely results. As we shall see in the next chapter there is a much simpler way of getting this. But even if it were the only way it would still not be necessary to calculate *all* the combinations.

An important property of the arithmetical triangle is that the series are symmetrical (see Table 1). The reason for this will be readily seen. If there are 4 possible combinations of 3 heads and 1 tail when 4 coins are thrown, then it must follow that there will be exactly the same number of combinations of 1 head and 3 tails. Taking 3 heads must be the same as taking 1 head for when combinations of 3 heads are made only one possibility of one head remains. After all, it is entirely a matter of point of view which we regard as a success—heads or tails. To the persons betting on the results of the throw, heads would be a success to one and tails to the other. So if there are $^4C_3$ ways of getting 3 successes (say, heads) on 4 coins thrown once, then this must be the same when from the other person's point of view 3 successes (now, tails) are the required result. And if 3 successes out of 4 are the same whether the successes are heads or tails, it must follow that $^4C_3 = {}^4C_1$. Moreover this is clearly a general rule and $^nC_r = {}^nC_{n-r}.$*

Because of this symmetry the sum of all the various combinations of n things could be obtained by calculating only half of these combinations. Moreover, if we had no table of the logarithms of factorials and were using the formula:

$$^nC_r = \frac{n(n-1)(n-2)\ldots(n-r+1)}{r!}$$

we could select the simpler calculation of the two. Instead of the very laborious calculation on page 17 for arriving at $^{100}C_{95}$, we should take $^{100}C_5$ which would give the same answer:

$$^{100}C_5 = \frac{100 \times 99 \times 98 \times 97 \times 96}{5 \times 4 \times 3 \times 2 \times 1}.$$

The question whether the results of throwing coins are combinations or permutations makes it important to spend a little more time considering permutations. On pages 9, 10 and 11 we were regarding the results in terms of succeeding or not succeeding as combinations. We asked: in how many different ways can, for example, three successes

---

* This can be proved also from the formulae:

$$^nC_r = \frac{n!}{(n-r)!\,r!}.$$

and

$$^nC_{n-r} = \frac{n!}{r!\,(n-r)!}.$$

But $(n-r)!\,r! = r!\,(n-r)!$

Therefore: $^nC_r = {}^nC_{n-r}.$

be obtained out of throwing 4 coins either once or in succession? Each success was regarded as being different. This would certainly be so if each coin were different, the first a half-crown, the second a florin, the third a shilling and the fourth a sixpence. A combination of heads on half-crown, florin and shilling would be quite distinct from heads on half-crown, shilling and sixpence. There are 4 possible combinations of 3 heads in a throw of 4 coins (or 4 throws of one coin):

$$H_1H_2H_3 \qquad H_1H_2H_4 \qquad H_1H_3H_4 \qquad H_2H_3H_4$$

The reasoning on pages 9, 10 and 11 is quite all right and the results can properly be called combinations.

Let us now look at the subject from a different point of view. We have two letters of the alphabet, T and H. In how many ways can these two letters be arranged if we have 3 of T and 1 of H and take all 4 at a time? We get the following answer:

$$H\,T\,T\,T \qquad T\,H\,T\,T \qquad T\,T\,H\,T \qquad T\,T\,T\,H$$

These are permutations and there are 4 possibilities. Had all 4 letters been different there would have been 4! (i.e. 24) permutations. There are not 4! because arrangements of the 3 Ts among themselves are all exactly the same, viz. TTT, so all those examples have to be eliminated. This means that, *each* time H is combined with different arrangements of the 3 Ts, the resulting permutations are identical: e.g. $H\,T_1\,T_2\,T_3$, $H\,T_2\,T_1\,T_3$, $H\,T_3\,T_2\,T_1$, $H\,T_3\,T_1\,T_2$, $H\,T_1\,T_3\,T_2$, $H\,T_2\,T_3\,T_1$, and the number of identical permutations is equal to the permutations of the identical letters. Thus, the number of permutations of 4 letters, of which three are alike, is $\frac{4!}{3!} = \frac{24}{6} = 4$.*

* There may, of course, be more than one letter duplicated. The letters in the word DEED can be arranged in 6 different ways:

DEED   DEDE   DDEE   EEDD   EDED   EDDE

Again, 4 different letters would have yielded 24 permutations. There are, in fact, only 6 permutations because the 2 Es and the 2 Ds each provide arrangements which are exactly the same. Thus the 2 letters E E can be followed by $D_1\,D_2$ or $D_2\,D_1$, but the result is the same: EEDD. Therefore, each time the two Es are combined with the two Ds we get identical arrangements equal to the number of permutations of the two Ds, viz. 2!

The same holds for the Es. Permutations of the two Es make no difference to the answer; the results are identical. Each time DD are combined with EE the permutations of the Es represent the number of identical permutations, e.g. $DDE_1\,E_2$, $DDE_2\,E_1$. In total there are, therefore, 2! multiplied by 2! permutations which will be exactly the same and the number of *distinct* permutations of 4 letters consisting of 2 of one and 2 of another is $\dfrac{4!}{2!\times2!} = 6$.

This is what happens when we arrange 2 heads and 2 tails. We get 6 permutations:

H H T T	H T H T	H T T H
T T H H	T H T H	T H H T

The general rule will now be clear. Where there are only 2 different things—e.g. success or failure, heads or tails, a six on a die or not a six—then if there are r of one in the group of n there can be only $n-r$ of the other and the permutations will be $\dfrac{n!}{r!(n-r)!}$. This formula is just the same as that given for combinations on page 17. They are, therefore, in one sense permutations and in another combinations. It is usual to refer to them as combinations, regarding them as the number of ways in which it is possible to get r successes in n events.

CHAPTER IV

THE ADDITION AND MULTIPLICATION PRINCIPLES

SO far we have dealt with chance in terms of throwing n coins once or one coin n times; but what happens if we toss n coins more than once? Will this mean that we have a better chance of getting a particular ^nC_r? On the principle of better luck next time, if there are enough throws of the n coins, ^nC_r is bound to appear sooner or later.

This sort of problem worried the Chevalier de Mere. In one of the games which he played, a die was cast 4 times and the betting by one player was that a 6 would appear at least once in these 4 throws and by the other player that it would not. The Chevalier discovered from long experience that there was a better chance of getting the 6 in the 4 throws than not getting it. It was a profitable bet.

A variation of this game was introduced. Instead of one die, two were thrown 24 times and the betting was for or against a double 6 appearing at least once in the 24 throws. Much to his surprise and cost, the Chevalier discovered that the better chance now was with the player who bet against the appearance of the double 6. He could not understand it. When one die is cast there are 6 possible results: when two dice are cast there are 36 possible results. As 24 to 36 is the same ratio as 4 to 6 surely the chance of getting a double 6 in 24 throws ought also to be better than not getting it. The chance should surely be the same as for getting a 6 in 4 throws of one die? Again the Chevalier wrote to Pascal, suggesting that, as the two chances were not consistent, arithmetic was not a reliable subject!

Pascal discussed this problem with Fermat and they solved it, but the method we should use today was worked out by the Swiss mathematician, Jacques Bernoulli (1654–1705), a contemporary of Isaac Newton. Bernoulli wrote a book, published eight years after his death, in which he discussed combinations and permutations, the solution of games of chance, and the manner in which probability might be made useful in social and economic affairs. It was he who introduced the word permutation, as we noted in the previous chapter. He dealt fully with the chance of succeeding a given number of times in a given number of throws.

The observant Chevalier clearly knew that the probability of getting a six was the ratio of the side containing the six to the total number of sides. He also knew that there were 36 alternative combinations of the pairs of dice, a fact which he could work out easily, for the number is manageable. What he did not realize was that the probability of $0.02\dot{7}$ (or $\frac{1}{36}$) applied only to the throwing of 2 dice once and on each subsequent toss the chance of getting a double six was still only $0.02\dot{7}$. Throwing 2 dice 24 times was similar to throwing one die 4 times *only* in the sense of regarding *all* the 24 throws as one set (just as 4 throws are regarded as one set). Tossing 4 dice once or tossing one die 4 times are similar because all the events are quite independent of one another and we regard the results of all 4 dice or all 4 throws together *as a group*.

When we consider the Chevalier's chance of getting a 6 at least once in 4 throws we must count the number of all the possible combinations of 4 containing one or more successes (obtaining a six being a success and not obtaining a six, i.e. getting any of the numbers from 1 to 5, as a failure). Similarly, when we seek the probability of obtaining a double six at least once in 24 throws of 2 dice, what we are really trying to find is the number of pairs of dice taken in sets of 24 pairs which contain a double six compared with the total number of combinations of all pairs of dice taken 24 at a time. In short, we regard each pair of results in exactly the same way as we regard one side of one die. A success is obtaining a double six; a failure is obtaining any other of the 26 pairs. There are $^{24}C_1 + ^{24}C_2 + ^{24}C_3 + \ldots + ^{24}C_{23} + ^{24}C_{24}$ ways of getting one or more double sixes.

There is, however, one important difference between getting a double six and getting a head on one coin—the chance of getting the double six is not 0.5 as with the coin but only $0.02\dot{7}$ ($\frac{1}{36}$). Thus we can hardly say that, because there are, for example, $^{24}C_3$ ways of getting 3 double sixes and 21 other pairs, these are equal to getting 24 double sixes. We should expect to get a higher probability of failures (not 6) combining than 6s because there are five times as many chances of not getting a 6.

Let us look at this more carefully. The probabilities of getting each side of a die are assumed to be equal. As there are six sides the probability of any one side being uppermost when the die is tossed is $0.1\dot{6}$ ($\frac{1}{6}$). With two dice we must begin with this equal probability of $0.1\dot{6}$ for each die and work out the number of equally likely chances of getting pairs of sides. The method is similar to that used for two coins on page 8 except that for each side on one of the dice we have 6 numbers on the other die with which it can be combined. Thus the side containing one spot on one of the dice—let us call it die A—can be combined with each of the sides containing 1, 2, 3, 4, 5 or 6 on die B:

A B	A B	A B	A B	A B	A B
1,1	1,2	1,3	1,4	1,5	1,6

Similarly the other sides on die A can each in turn be combined with the 6 sides on die B:

2,1	2,2	2,3	2,4	2,5	2,6
3,1	3,2	3,3	3,4	3,5	3,6
4,1	4,2	4,3	4,4	4,5	4,6
5,1	5,2	5,3	5,4	5,5	5,6
6,1	6,2	6,3	6,4	6,5	6,6

If *each* of 6 sides of one thing can be combined with *all* of 6 sides of another it follows that there must be $6 \times 6 = 6^2 = 36$ alternative equally probable pairs.

If now we take 3 dice instead of 2 we shall find that every one of the 36 pairs can be combined with all 6 sides of the third die. Thus all 36 pairs can be combined with the side of die C marked with one spot:

A B C	A B C	A B C	A B C	A B C	A B C
1,1,1	1,2,1	1,3,1	1,4,1	1,5,1	1,6,1
2,1,1	2,2,1	2,3,1	2,4,1	2,5,1	2,6,1
3,1,1	3,2,1	3,3,1	3,4,1	3,5,1	3,6,1
4,1,1	4,2,1	4,3,1	4,4,1	4,5,1	4,6,1
5,1,1	5,2,1	5,3,1	5,4,1	5,5,1	5,6,1
6,1,1	6,2,1	6,3,1	6,4,1	6,5,1	6,6,1

This can be repeated with the side of die C containing 2 spots, then with the 3 spots side, and so through all 6. The total number of equally likely combinations of the results of throwing 3 dice is therefore $36 \times 6 = 6^3 = 216$.

With 4 dice the total number of equally likely combinations of 4 sides would be 6^4; with 5 dice 6^5; with 10 dice 6^{10}; and so on. Thus the number of combinations of n things which are equally likely to appear is derived from the number of equally likely alternatives of each multiplied n times. If the number of equally likely alternatives on each is x, the total combinations of n in sets of n will be x^n.

The number of equally likely alternatives of each is not always identical but the multiplication rule still applies. If, for example, a die were tossed simultaneously with a coin, the number of pairs would be equal to the 6 equally likely results from the die multiplied by the 2 equally likely results from the coin, i.e. 12, because for every side of the die it would be possible to get either a head or a tail, or, conversely, for each side of the coin it would be possible to combine any of the sides of the die from 1 to 6.

This multiplication rule is important and, taken in conjunction with the rule of addition, forms the basis of probability calculations. Before proceeding it will be advisable to state these rules formally.

Let us begin with the rule relating to addition. In a given set of equally likely results the probability of getting results having a particular attribute is the sum of the probabilities of all such results. If there are N possible alternative results all equally likely the probability of getting any one is $\dfrac{1}{N}$ and the probability of getting a particular attribute found in m of the results is $\dfrac{m}{N}$. Thus the probability of drawing any one card from an ordinary pack of 52 playing cards is $\frac{1}{52}$ so the probability of getting an ace is the sum of the probabilities of getting each of the four aces in the pack, viz. $\frac{1}{52}+\frac{1}{52}+\frac{1}{52}+\frac{1}{52} = \frac{4}{52} = 0\cdot07692$. The probability of drawing an ace, king, queen or jack is $\frac{16}{52} = 0\cdot30769$, made up of the sum of the probabilities of getting each separately. On a die the probability of getting any one of the equally likely sides is $\frac{1}{6}$, so the probability of getting an even number is the sum of getting either 2 or 4 or 6, i.e. $\frac{3}{6} = 0\cdot5$.

This is known as *the principle of addition of probabilities*, which may be stated as follows: If in a set of N equally likely, independent and mutually exclusive results m_1 are favourable to getting an attribute X, another m_2 to getting attribute Y, and the remaining m_3 to getting attribute Z, then the probability p_1 of getting X is $\dfrac{m_1}{N}$, the probability p_2 of getting Y is $\dfrac{m_2}{N}$, and the probability p_3 of getting Z is $\dfrac{m_3}{N}$. The probability of getting either X or Y is p_1+p_2. As, however, $m_1+m_2+m_3 = N$, it follows that $p_1+p_2+p_3 = 1$ and the probability p_{1+2} of getting either X or Y may be written thus: $p_{1+2} = 1-p_3$.

The conditions that the results must be independent and mutually exclusive are important. By independent is meant that the probability of getting one result is not affected by the occurrence of any of the other alternative results. Thus the probability of getting a 6 when a die is thrown is not influenced by the fact that the last three throws have produced sixes each time—although many people wrongly think that it is. Whatever the previous results the probability of getting a six uppermost is always $0\cdot1\dot6$. On the other hand, if we propose drawing blindfolded 3 balls in succession and without replacement from an urn containing 60 red and 40 white balls the probability that the second will be red depends on whether the first result was a red or white ball. If the first ball was red, the probability that the second will be red is $\frac{60-1}{100-1} = 0.5\dot{9}$; if the first ball was white, the probability that the second will be red becomes $\frac{60}{100-1} = 0.6\dot{0}$. The probability that the third ball will be red is similarly dependent on the previous two results.

First result	Second result	Probability that third result will be red
Red	Red	$\frac{60-2}{100-2} = 0.59184$
Red	White	$\frac{60-1}{100-2} = 0.60204$
White	Red	$\frac{60-1}{100-2} = 0.60204$
White	White	$\frac{60}{100-2} = 0.61224$

In these circumstances the results are not independent and the probabilities, after the first drawing, are dependent on the previous results. We say that 0.59 is the *conditional probability* of getting red second given that red was first. The probabilities would, however, cease to be conditional if each ball was replaced before making another withdrawal. Whether a red or white ball had been withdrawn first would then make no difference to the second result and the probability of getting red would always be 0.6; the results would be independent of one another.

By mutually exclusive is meant that the occurrence of one result makes the simultaneous occurrence of any of the others impossible. If we have drawn one card—a seven of clubs—we cannot have any of the other 51 possible results. The property of being mutually exclusive is clearly essential for the addition principle to be valid. If we have 20 discs numbered from 1 to 20 the probability of getting a number divisible by 4 would be the addition of the separate probabilities of getting discs numbered 4, 8, 12, 16 or 20, i.e. 0.25 and the probability of getting numbers divisible by 5 would be the addition of the separate probabilities of getting the discs numbered 5, 10, 15 or 20 which is 0.20. It would, however, be wrong to say that the probability of getting a number divisible by either 4 or 5 is 0.25+0.20 because getting a number divisible by 4 does not exclude the chance of getting a number divisible by 5.

The principle of multiplication of probabilities relates to probabilities in *different* sets of events. There are six sides to a die and as the appearance of each is assumed to be equally likely the probability of getting each is 0.16 $(\frac{1}{6})$; there are two sides of a coin so the probability of getting one of those sides is 0.5 $(\frac{1}{2})$. When, however, two dice are taken together we get a *new* set of 36 pairs of sides and the probability of getting both one side of die A and one side of die B is $\frac{1}{6} \times \frac{1}{6} = \frac{1}{36}$ or 0.027. Similarly the appearance of both a side of a die and a face of a coin when the results of throwing a die and coin are taken together yield 12 new events and the probability of getting each event is $\frac{1}{6} \times \frac{1}{2} = \frac{1}{12} = 0.083$.

In general if p_a is the probability of getting a particular attribute in set A, p_b the probability in another set B, p_c in set C, and so on, the probability of getting all these attributes together in a new set derived from the three other sets is $p_a \times p_b \times p_c$ which can also be written $p_a \cdot p_b \cdot p_c$. (The dot means 'multiplied by'.)

A probability derived from the multiplication principle creates and forms part of a *new* set. The addition principle relates always to probabilities in the *same* set of events. Both rules make possible the computation of probabilities without having to list all the alternative equally likely combinations.

We apply the principle of addition of probabilities to the new derived set when we think in terms of two attributes, success or failure. These may or may not be equally likely. If, in a new set of two dice results, we want to get pairs consisting entirely of even numbers (2,2; 2,4; 6,2; etc.) we must first of all find the probability of getting even numbers in each of the two sets from which the pairs are to be formed. It will be readily seen that these probabilities are $\frac{1}{2}$ for each die—3 even numbers on the 6 sides of each. Therefore, the probability of getting *both* even numbers in the pairs of the derived set is $\frac{1}{2} \times \frac{1}{2} = \frac{1}{4}$. Reference to the list of pairs on page 23 shows that there are 9 combinations of even numbers out of the total of 36 pairs, which gives the same chance of $\frac{1}{4}$. Similarly, the chance of getting pairs of odd numbers is the multiplication of the probabilities of getting odd numbers on each die. Again this is $\frac{1}{2} \times \frac{1}{2} = \frac{1}{4}$.

Because the chances of success and of failure are equal in the initial sets we are just as likely to get an odd number combining with an even as we are to get 2 evens. But we must be careful. We can have an even on die A with an odd number on die B, or, conversely, we can have odd on A with even on B. In the language of combinations there are

2 ways of getting one success (an even) in combinations of two, $^2C_1$. The probability of 0·25 relates to getting *one* combination only, e.g. getting an even on die A and an odd on die B. Therefore the probability* of getting all combinations of odd and even numbers is equivalent to the additions of the probabilities in the new set of getting each of these combinations, i.e. $\frac{1}{4} + \frac{1}{4} = \frac{1}{2}$.

However, it doesn't always work out like this. Let us now examine the chance of getting at least one 6 in the pairs obtained from tossing 2 dice. We already know from page 23 that the answer is 11 such pairs out of a possible total of 36 equally likely pairs—a probability of 0·30$\dot{5}$.

We start by applying the addition principle. The probability of getting a 6 on die A is $\frac{1}{6}$ and of not getting a 6 must be the sum of the probabilities of getting 1, 2, 3, 4 and 5, i.e. $\frac{1}{6} \times 5$ or $1 - \frac{1}{6} = \frac{5}{6}$; on die B the probabilities are similar.

Next, we apply the multiplication principle. The chance of getting a double 6 is $\frac{1}{6} \times \frac{1}{6} = \frac{1}{36}$; the chance of getting one 6 and any other number is $\frac{1}{6} \times \frac{5}{6} = \frac{5}{36}$; and the chance of obtaining pairs with no 6 at all is $\frac{5}{6} \times \frac{5}{6} = \frac{25}{36}$. In these circumstances we clearly cannot take the frequency of combinations alone, $^2C_2$, $^2C_1$ and $^2C_0$, as a measure of probability for each combination is not equally likely. The probabilities of getting each combination—$\frac{1}{36}$, $\frac{5}{36}$ and $\frac{25}{36}$ respectively—are important. It will be noticed that the probabilities of all three possible combinations certainly do not add up to $\frac{36}{36}$.

The next step is to apply the principle of addition to the new set of 36 pairs. There is only one combination representing a double 6 and only one representing no 6s, but there are 2 ways of getting a combination of one 6 and another number ($^2C_1$). The success can be on either die A or die B. So we add the probability of getting each of those combinations together: $\frac{5}{36} + \frac{5}{36} = \frac{10}{36}$.

We now have the answer. The chance of getting at least one 6 in a

* In tabular form this can be restated as follows:

Combinations of 2 dice	Probability of each combination	Number of combinations	Total probabilities $(2) \times (3)$
(1)	(2)	(3)	(4)
2 evens	$\frac{1}{2} \times \frac{1}{2} = \frac{1}{4}$	$^2C_2 = 1$	$\frac{1}{4} = 0·25$
1 even and 1 odd	$\frac{1}{2} \times \frac{1}{2} = \frac{1}{4}$	$^2C_1 = 2$	$\frac{1}{2} = 0·50$
No evens and 2 odds	$\frac{1}{2} \times \frac{1}{2} = \frac{1}{4}$	$^2C_0 = 1$	$\frac{1}{4} = 0·25$
		4	1·00

What we have found are 4 combinations each having a probability of 0·25. When the probabilities of each result are equal, the frequency of the combinations alone will give the required probabilities as shown above and for the coins on pages 7 and 8.

toss of two dice is $\frac{1}{36}+\frac{10}{36} = \frac{11}{36}$. This agrees with our previous knowledge. Moreover, all the results now add up to unity: $\frac{1}{36}+\frac{10}{36}+\frac{25}{36} = \frac{36}{36}$.

All this can be generalized. If p is the proportion of individuals in a
set having a required attribute, then the proportion not having that
attribute must be $1-p$ which may be represented by q, i.e. $p+q = 1$.
When more than one individual drawing is made either by repeated
trials from the same set or by combinations of individuals drawn from
identical sets, the probability of getting any one combination containing
all n successes will be p^n, all failures q^n, and r successes $p^r q^{n-r}$. There is,
however, more than one way of getting a combination containing r
individuals with the required attribute so the total chance of getting r
successes is $^nC_r p^r q^{n-r}$.

If $p = q = \frac{1}{2}$, then $p^n = q^n = p^r q^{n-r}$, and the probability of getting r

successes is measured simply by $\dfrac{^nC_r}{^nC_0 + {^nC_1} + \ldots + {^nC_{n-1}} + {^nC_n}}$. But,

in those circumstances $(\frac{1}{2})^n$ will be equivalent to $1/^nC_0 + {^nC_1} + \ldots +$
$^nC_{n-1} + {^nC_n}$. The total of $^nC_0 + {^nC_1} + {^nC_2} + {^nC_3} + \ldots + {^nC_{n-1}} + {^nC_n}$ can

be more conveniently written as $\displaystyle\sum_{r=0}^{r=n} {^nC_r}.$*

We have thus found a quick method of arriving at the total number
of all possible ways of getting r successes in n events:

$$\sum_{r=0}^{r=n} {^nC_r} = 2^n.\dagger$$

* In mathematics, the Greek capital S (Σ, pronounced sigma) is used to signify
'the sum of'; thus, $\displaystyle\sum_{r=0}^{r=n} {^nC_r}$ means 'the sum of all combinations of n events, having no
successes and all numbers of successes up to all n successes'. So $(\frac{1}{2})^n = 1/\displaystyle\sum_{r=0}^{r=n} {^nC_r}$.

† This can be seen from the arithmetical triangle, for each row is a doubling of
the row above it, thus:

1							1
	1						
1	1					2	2
	1	1				+2	
1	2	1				4	$2 \times 2 = 4 = 2^2$
	1	2	1			+4	
1	3	3	1			8	$4 \times 2 = 8 = 2^3$
	1	3	3	1		+8	
1	4	6	4	1		16	$8 \times 2 = 16 = 2^4$
	1	4	6	4	1	+16	
1	5	10	10	5	1	32	$16 \times 2 = 32 = 2^5$

The expression 2^n gives the total number of all combinations of n items. When $p = \frac{1}{2}$, the probability of getting each separate combination is $(\frac{1}{2})^n$ and the probability of getting all the combinations would be $2^n \times (\frac{1}{2})^n = \dfrac{2^n}{2^n} = 1$. In other words we are clearly certain to get one or other of all possible combinations and certainty is measured by unity.

In trying to find the probability of obtaining 95 heads and 5 tails in a throw of 100 coins we found on page 17 that it could occur in 75,280,000 different ways. We now know that the total number of all combinations of heads and tails on 100 coins is 2^{100}. The probability of getting 95 heads and 5 tails is therefore $\dfrac{^{100}C_{95}}{2^{100}}$. This can be easily worked out with the aid of logarithms:

(1) Log 2   ·3010300
(2) Log 2 multiplied by 100   30·1030000
(3) Log $^{100}C_{95}$ (see page 17)   7·8767
(4) Deduct (2) from (3) $\overline{23}$·7737
(5) Anti-log of $\overline{23}$·7737   0·00000000000000000000005939

In short it is very unlikely!

Let us now return to the Chevalier's problem. He could not understand why the probability of getting at least one 6 in 4 throws of a die was not the same as getting a double 6 at least once in 24 throws of 2 dice. His approach to the problem was quite wrong. What we are really asking is: what are the chances of getting in combinations of 4 throws of a die at least one success (i.e. a 6); what also are the chances of getting in combinations of 24 throws of 2 dice at least one pair which will show a 6 on the uppermost side of each. The initial probability from which we begin is not $\frac{1}{2}$ but $\frac{1}{6}$ so the probability of getting each combination is not the same for all combinations. We must use the formula $^nC_r p^r q^{n-r}$.

The chance of getting a 6 in the first of the 4 throws and some other number on the second, third and fourth throws is $(\frac{1}{6}) \times (\frac{5}{6})^3 = \frac{1}{6} \times \frac{125}{216} = \frac{125}{1296}$. However, the success might be obtained on the second throw instead of the first, or on the third or fourth, because there are 4 ways in which it is possible to have one success in 4 throws ($^4C_1$). Therefore the probability of getting one 6 in 4 throws is $\frac{125}{1296} + \frac{125}{1296} + \frac{125}{1296} + \frac{125}{1296} = \frac{125}{1296} \times 4 = \frac{500}{1296} = 0·385803$.

This is not the probability of winning the game, for a player might get not just one 6 but 2 or even 3 or 4. The chance of winning the game to the player betting, as apparently the Chevalier de Mere did, on at least one 6 appearing in 4 throws of a die, is:

$$^4C_1pq^3 + {}^4C_2p^2q^2 + {}^4C_3p^3q + {}^4C_4p^4.$$

Thus we want the probability of getting all results except no sixes. The probability of getting no 6s is $^4C_0q^4 = 1 \times (\frac{5}{6})^4 = 0.482253$. Therefore, the probability of getting all other results is $1 - 0.482253 = 0.517747$, i.e. just over 0.5 as the Chevalier found from his experience.

In dealing with the problem of getting a double six at least once in 24 throws we begin with an initial probability of $\frac{1}{36}$. A success is getting a pair containing two sixes and as there can be only one such pair out of a total of 36 pairs then $p = \frac{1}{36}$ and $q = 1 - \frac{1}{36} = \frac{35}{36}$. Remember that we are taking the 24 throws as a *new* set of events. We are asking: what is the probability of getting 24 pairs containing one double 6 and 23 other pairs. If we get the double six last of all preceded by 23 failures the probability will be $\frac{1}{36} \times (\frac{35}{36})^{23} = 0.014532$. But there are 24 ways of getting 1 double 6 and 23 failures ($^{24}C_1$) so the total probability of getting just 1 double 6 is 0.014532 multiplied by 24 = 0.348752 (i.e. the addition of the probabilities of all 24 mutually exclusive results). Similarly we can calculate the chance of getting 2 double sixes in 24 throws, then 3, and so on. Again we want any result except 24 failures (no double sixes), the probability of which is $^{24}C_0q^{24} = 1 \times (\frac{35}{36})^{24} = 0.508596$. The probability of succeeding in getting a double 6, once or more often, in 24 throws is thus $1 - 0.508596 = 0.491404$. As this is less than 0.5 it confirms the Chevalier's observation that the person who bet against its appearance won more often than he who bet for it. And, of course, the experience confirms the mathematics and does not, as the Chevalier told Pascal, prove it to be unreliable!

Chapter V

FINDING THE UNKNOWN

ALTHOUGH James Bernoulli devoted a large part of his book *Ars Conjectandi* to a discussion of games of chance, he was aware that the knowledge might be of value in other ways and, in the fourth part, he dealt with the application of these ideas to civil and economic affairs. He died before completing that part, but in this book he left ideas which are the kernel of modern sampling theory.

The formula for finding the probability of getting r successes in n tries was given in the previous chapter as $^nC_rp^rq^{n-r}$. Bernoulli noted that all the possible results obtainable from giving to r values ranging from 0 to n could be obtained from an expansion of the initial probability of success (p) and of failure (q) by the binomial theorem. This theorem gives a ready answer to the multiplication of any two terms, say $q+p$ by themselves n times, i.e. $(q+p)^n$. When this is expanded for $n = 2$, $n = 3$, $n = 4$, $n = 5$, we get the following interesting answer:*

$$* (q+p)^2 = (q+p)(q+p)$$
$$= q+p$$
$$\underline{\quad q+p\quad}$$
$$q^2+qp$$
$$\underline{\quad pq+p^2\quad}$$
$$q^2+2qp+p^2$$

$$(q+p)^3 = (q+p)(q+p)(q+p)$$
$$= (q^2+2qp+p^2)\times(q+p)$$
$$= q^2+2qp+p^2$$
$$\underline{\qquad q+p\qquad}$$
$$q^3+2q^2p+qp^2$$
$$\underline{\quad q^2p+2qp^3+p^3\quad}$$
$$q^3+3q^2p+3qp^2+p^3$$

$$(q+p)^4 = (q+p)(q+p)(q+p)(q+p)$$
$$= (q^3+2q^2p+3qp^2+p^3)\times(q+p)$$
$$= q^3+3q^2p+3qp^2+p^3$$
$$\underline{\qquad q\ +p\qquad}$$
$$q^4+3q^3p+3q^2p^2+qp^3$$
$$\underline{\quad q^3p+3q^2p^2+3qp^3+p^4\quad}$$
$$q^4+4q^3p+6q^2p^2+4qp^3+p^4$$

$$(q+p)^5 = (q+p)(q+p)(q+p)(q+p)(q+p)$$
$$= (q^4+4q^3p+6q^2p^2+4qp^3+p^4)\times(q+p)$$
$$= q^4+4q^3p+6q^2p^2+4qp^3+p^4$$
$$\underline{\qquad q\ +p\qquad}$$
$$q^5+4q^4p+6q^3p^2+4q^2p^3+qp^4$$
$$\underline{\quad q^4p+4q^3p^2+6q^2p^3+4qp^4+p^5\quad}$$
$$q^5+5q^4p+10q^3p^2+10q^2p^3+5qp^4+p^5$$

$$(q+p)^2 = q^2 + 2qp + p^2$$
$$(q+p)^3 = q^3 + 3q^2p + 3qp^2 + p^3$$
$$(q+p)^4 = q^4 + 4q^3p + 6q^2p^2 + 4qp^3 + p^4$$
$$(q+p)^5 = q^5 + 5q^4p + 10q^3p^2 + 10q^2p^3 + 5qp^4 + p^5)$$

The reader will recognize these answers. They are the series of numbers from the arithmetical triangle multiplied by the probability of getting each combination. For example, $10q^3p^2$ is simply $^5C_2q^3p^2$, and $4q^3p$ is $^4C_1q^3p$. Therefore, the expansion of $(q+p)^n$ gives all the terms of $^nC_rq^{n-r}p^r$ from $^nC_0q^n$ to $^nC_np^n$, i.e. through all the values of r from $r = 0$ to $r = n$. The explanation is quite simple. One term in each pair has to be multiplied by one term in each of the other pairs. Thus, $(q+p)^3$ is $(q+p) \times (q+p) \times (q+p)$. This is the same as finding the number of ways in which heads and tails on three coins can be combined. Each bracket is like a coin with q as one side and p as the other.

The answer* is the same as we previously had of—

$$q^3 + 3q^2p + 3qp^2 + p^3.$$

There is no magic in the probability distribution which results from expansion of the binomial, $(q+p)^n$, because the multiplication process merely involves combining one term in each bracket with a term in every other bracket. Hence, there will be ^nC_r ways in which p can be multiplied by itself r times and, whenever r is smaller than n, this means that there will be ^nC_r ways in which multiplications will give $q^{n-r}p^r$.

This distribution will be symmetrical when $q = p = \frac{1}{2}$ because both $q^{n-r}p^r$ and q^rp^{n-r} equal $(\frac{1}{2})^n$ and ^nC_r equals $^nC_{n-r}$ (see Tables 3, 4 and 5); the distribution will be asymmetrical for other values of q and p because

---

* First, let us multiply (i.e. combine) the three qs:

$$(q+p)(q+p)(q+p) = q^3 \qquad \text{(one combination).}$$

Each q has also to be multiplied by each p in other brackets:

$$(q+p)(q+p)(q+p) = 3qp^2 \qquad \text{(three ways of combining one failure (q) with two successes ($p \times p$)).}$$

Next, p can be multiplied first by each p and then by each q in other brackets:

$$(q+p)(q+p)(q+p) = p^3 \qquad \text{(one combination)}$$

$$(q+p)(q+p)(q+p) = 3q^2p \qquad \text{(three combinations of two failures ($q \times q$) with one success, p).}$$

As there are no other combinations possible we have done all the multiplications.

in those circumstances $q^{n-r}p^r$ will not equal $q^r p^{n-r}$ and $^nC_r q^{n-r}p^r$ will not, in consequence, equal $^nC_{n-r}q^r p^{n-r}$ (see Tables 6, 7 and 8). In statistics, the first is known as the *symmetrical binomial frequency distribution* and the second as the *asymmetrical binomial frequency distribution*.

Bernoulli asked an important question. Assume that we do not know the initial probability of getting a given attribute—for example, heads—can it be inferred from the results obtained? The answer to this question is not only of considerable practical significance but also of theoretical interest. So far we have assumed that the probability of getting a head or a tail in tossing a coin is 0·5 simply because our minds accept such an assumption as reasonable. We cannot think of any other value which it might be. Similarly, with a six-sided die we assume that every side is as likely as the others to be face uppermost in a long series of throws and therefore we make the probability of getting each side exactly one-sixth, i.e. 0·16. Yet it could be argued that, as six dots have to be cut out of one side and a smaller number of cavities out of the others, the die is not so exactly balanced on each side and the probabilities may not, in fact, be equal. Moreover, although in problems concerning dice or coins or cards we can make what appears to be a reasonable assumption about an initial probability from which we can compute other probabilities, this is rarely possible in general statistical work.

It is necessary at this point to examine more closely our definition of probability. Throughout the book we have defined probability as the proportion represented by the number of possible events containing a desired attribute out of the number of all possible equally likely events in the same set. This is the definition adopted by the mathematicians who developed what is usually called the classical theory of probability. Thus, when tossing one coin we say that the probability of getting a head is 0·5 simply because there are only two possible results, a tail or a head, and we assume each to have equal probabilities of turning up. When tossing two coins we say the probability of getting two heads is 0·25 because there are only four possible combinations of heads and tails, only one of which consists of two heads and each combination is considered to be equally likely. These probabilities are, however, based on assumptions of probabilities—the assumption that the sides of each coin have equal probabilities.

What does this definition mean? It can only be that events which are assumed to be equally likely will, in an actual set of results, appear an equal number of times. If this is not so, the definition can have no meaning. How can we speak of the chance of getting heads as being equal to the chance of getting tails in throws of a coin if the results consistently favour one or the other? Our definition of probability must imply that heads will appear no more and no less often than tails in the results obtained from throwing a coin.

Objection may be taken to defining probability in terms of itself. We ought to define equally likely events. If, however, we say that they are events which occur with equal frequency in an actual number of results, we have to define the circumstances in which this happens. How many times is it necessary to toss a coin to be quite certain that exactly half the total number of results will be heads? Only when we know this can we estimate probabilities from results and when we know this have we not an entirely new definition of probability based on relative frequencies and not dependent on the assumption of equally probable events?

Here we get the beginnings of sampling. A sample is a portion which serves as an example of the whole; the word is derived from an old English word, *essample*, meaning an example. What we are looking for is a number of results—a sample—which will give an accurate estimate of the initial probability of getting a particular attribute where the probability cannot be known beforehand. If we do not know beforehand, we can, of course, never check our estimate. To find this sample, we are forced to assume that we do know the probabilities beforehand. We must, therefore, reason from the knowledge so far obtained and continue to accept initial equal probabilities as valid. Although the assumption can be challenged, its origin in games of chance makes it seem reasonable enough and it has produced solutions which have been confirmed by gamblers' experiences, as we have seen in previous chapters. Thus, there is already some evidence that the early mathematicians' hypotheses are correct. Other evidence has since been collected but the classical ideas can provide a basis for reasoning about the size of sample necessary to enable us to use results for estimating probabilities.

Let us take a sample of 100 throws of a coin (or one throw of 100 coins). In order to obtain a true estimate of the initial probabilities, q and p, we require a result showing 50 heads and 50 tails. This may be stated as follows:

$$100(q+p)$$
$$= 100(0 \cdot 5 \text{ T} + 0 \cdot 5 \text{ H})$$
$$= 50 \text{ tails} + 50 \text{ heads.}$$

But we immediately realize that this is only one combination out of various possible combinations of 100 coins or throws, from 100 tails and no heads to no tails and 100 heads ($^{100}C_0$ to $^{100}C_{100}$). The probabilities of getting each of these combinations can be found from an expansion of the binomial: $(q+p)^{100}$. Thus, the probability of getting exactly 50 tails and 50 heads is $^{100}C_{50}(\frac{1}{2})^{50}(\frac{1}{2})^{50} = 0.07849$.

We now have a new probability. Again, if this probability has any meaning it must be that 0.07849 of actual results will consist of combinations of 50 tails with 50 heads. If we take a sample of 1000 throws of 100 coins (or 100,000 throws of one coin in groups of 100 successive throws), we should expect to get 78 events (0.07849×1000) which were exactly 50 tails and 50 heads. But, to be consistent, all the other possible combinations should be found occurring with the frequencies indicated by the theoretical probabilities which can be calculated from $^{100}C_r q^{100-r} p^r$. In other words, the actual frequencies of the different possible combinations of 100 coins should be in accordance with the expansion of:

$$1000(q+p)^{100}.$$

Our interest, of course, is in getting results which will give a true estimate of the initial probability of getting heads. If our assumption of the equality of probabilities of getting heads and tails is correct, we must get a sample showing an equal number of heads and tails if our estimate of the initial probability is to be correct. We are not, therefore, interested in getting the other combinations. If, however, we obtained a sample of 1000 sets of 100 results agreeing with the expansion of $1000 (q+p)^{100}$ we should, in fact, have a sample consisting of 50,000 tails and 50,000 heads. The reason is obvious when our initial probability is 0.5 for this means that $q^{n-r}p^r = q^r p^{n-r} = \frac{1}{2}^n$, and, as $^nC_r = {}^nC_{n-r}$, then $^nC_r q^{n-r}p^r = {}^nC_{n-r}q^r p^{n-r}$. If we add a combination of 7 tails and 93 heads to one of 93 tails and 7 heads we get 100 tails and 100 heads. As there are an equal number of each opposite combination we always get an equal number of tails and heads in our total results. Similarly, in an asymmetrical distribution the total number of successes expressed as a proportion of total results will be just the same as p; or, to state this truth differently, the average number of successes (r) in a binomial distribution is always equal to np. This is demonstrated in Table 2.

Again, however, we know that a sample consisting of 50,000 tails and 50,000 heads is only one out of a number of different combinations possible in 100,000 throws of a coin. We might get 49,802 tails and

50,198 heads, or any combination from 100,000 tails and no heads to 100,000 heads and no tails. ($^{100,000}C_0$ to $^{100,000}C_{100,000}$.) The probabilities of getting all these are to be found from an expansion of the $(q+p)^{100,000}$. The probability of getting exactly 50,000 tails and 50,000 heads is thus $^{100,000}C\,(\frac{1}{2})^{50,000}(\frac{1}{2})^{50,000} = 0\cdot002523$.

This new probability must mean that $0\cdot002523$ of the results of throwing 100,000 coins (or one coin in groups of 100,000 successive throws) will consist of combinations of exactly 50,000 tails and 50,000 heads. Again, we may ask: how many results (how many sets of 100,000 throws) is it essential to have to be sure of this? Assume that we could take a sample of one million such combinations of 100,000 throws of a coin. In this sample we expect to get 2523 combinations with 50,000 tails and 50,000 heads (i.e. $0\cdot002523$ multiplied by 1 million). But we ought also to get all the other possible combinations with the frequencies indicated by the expansion of the binomial $(q+p)^{100,000}$. These frequencies will be:

$$1,000,000\ (q+p)^{100,000}.$$

If this is so our sample of 100,000,000,000 results will, in fact, contain exactly 50,000 million tails and 50,000 million heads.

But this, of course, is only one combination of 100,000,000,000 throws (or coins) out of all the different possible combinations from 100,000,000,000 tails and no heads to no tails and 100,000,000,000 heads . . .

We can go on and on with this reasoning. Every probability involves a new probability; we are never certain of getting a sample which accurately reflects the initial probability. This is a rather disturbing result. There appears to be no sample, however large, which will accurately give the initial probability with certainty. We can never be sure that an initial probability inferred from results will be correct. Indeed, the probability of getting an exact estimate seems to diminish as the size of the sample increases. This is not at all what we should intuitively expect. The smaller the sample, the greater is the chance of getting extremes. Surely, we are more likely to get 10 heads in a row than 100,000 heads in a row; a large sample ought to give a better estimate of the initial probability than a small one.

Bernoulli was well aware of these difficulties; in particular, the impossibility of getting a certain answer. He therefore set himself the problem of finding the limits within which the initial probability must lie. As the probability of getting extreme proportions and of getting a proportion exactly equal to the initial probability both decrease as the size of the sample increases, the probability of getting a good estimate *within stated margins of error* should increase as the size of the sample increases.

We shall follow this argument most clearly with the aid of diagrams. Let us examine a number of expansions of $(q+p)^n$. Tables 3, 4 and 5 show the expansion of $(q+p)^n$ where $q = \frac{1}{2}$, $p = \frac{1}{2}$ and n has the values 10, 24 and 100. We have here three frequency distributions indicating the probabilities of getting every possible answer when we take a sample of 10, or of 24, or of 100. There are two ways of looking at the results: either we have a number of heads or a proportion of heads in our sample. Each table therefore contains two frequency distributions. There is the frequency (i.e. probability) of getting a given *number* of successes (e.g. heads) and there is the frequency of getting a given *proportion* of successes.

Chart 2 shows the three frequency distributions for getting a *number* of heads. The probability of getting each number of heads is represented by an area; the sum of all these areas thus equals 1·0. Because the columns in each diagram in Chart 2 have equal width bases, the total areas in each diagram are equal to one another.

In Chart 3 the same three distributions are shown in terms of the *proportions* of heads in each sample. These are especially interesting, for they represent the estimates of $p = \dfrac{r}{n}$ from all the samples of size n.

Once again the probabilities are represented by areas but the total areas in each diagram are not now equal. The reason for this is clear. The length of the horizontal scale (which is known as the 'abscissa' by mathematicians) has been made identical in each diagram and the conversion of numbers into proportions has therefore resulted in a corresponding contraction in the width of the base of the columns as the samples increase in size. Thus, because the abscissa scale in Chart 3 for $(\frac{1}{2}+\frac{1}{2})^{24}$ is the same as for $(\frac{1}{2}+\frac{1}{2})^{10}$ instead of being 2·27 times (i.e. $\frac{25}{11}$) longer, as it is in Chart 2, the base of each column is only $\frac{11}{25}$ and the total area is also $\frac{11}{25}$ of the area of the chart for $(\frac{1}{2}+\frac{1}{2})^{10}$. Similarly as the abscissa scale for samples of 100 is identical to the others instead of being $\frac{101}{11}$ times longer than for samples of 10 and $\frac{101}{25}$ times longer than for samples of 24, the widths of each column in the diagram for samples of 100 must be $\frac{11}{101}$ and $\frac{25}{101}$ of the widths of the columns and $\frac{11}{101}$ and $\frac{25}{101}$ of the total areas in the other two diagrams. In order to make the total areas in each diagram the same, yet retain the identical abscissa scales, it is necessary to lengthen the vertical scales (known as 'ordinates'). If we halve the base of a column, we must double its height in order to keep the area the same, thus:

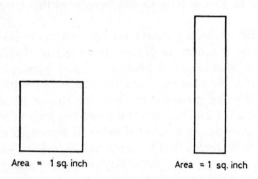

Area = 1 sq. inch   Area = 1 sq. inch

If we reduce the base of each column by $\frac{11}{25}$ we must increase their height by $\frac{25}{11}$ if we wish to keep the areas equal to one another; and so on. Unfortunately this process of raising the heights by stretching the ordinate makes the diagrams for large samples very tall. The effect of doing this to Chart 3 will, therefore, be left to the reader's imagination. As it stands, the chart makes very clear the narrowing of the total area which results from taking larger samples. In other words, as the sample increases in size so does the probability of getting a proportion of heads within stated limits of the true initial probability.

Let us see what the probability is of getting a result anywhere from 0·4 to 0·6 heads. The chances can be calculated from Tables 3, 4 and 5. They are 0·6562 for samples of 10, 0·6926 for samples of 24, and 0·9646 for samples of 100. Therefore we should, in 9646 out of 10,000 samples of 100 get an estimated initial probability of from 0·4 to 0·6.

This confirms what we felt in our bones; the bigger the sample, the better is our chance of getting *near* the true initial probability. We get nearer to it without ever reaching it. A number which is approached in this way is what mathematicians call a 'limiting value'—a concept invented in order to make practicable use of the idea of infinity, which so baffled mathematicians before the seventeenth century.

In Tables 6, 7 and 8 and in Charts 4 and 5 the expansions of $(q+p)^n$ are shown where $p = \frac{1}{6}$ (0·1$\dot{6}$), $q = \frac{5}{6}$ (0·83) and n has the values 10, 24 and 100 as before. Once more the probabilities are represented by areas and the total areas in each diagram are equal in Chart 4 but not in Chart 5. Again we note that the curve of proportions (the estimates of p) narrows as the size of sample is increased. It is also noticeable that the curves in both charts become less asymmetrical with larger samples.

Let us list the conclusions which are suggested by these diagrams:

(i) When $p = q = 0.5$, the distributions are symmetrical for all values of n, i.e. for samples of all sizes.

(ii) When p is not equal to q, the distributions are asymmetrical; they have what statisticians term 'skew'.

(iii) This 'skew' diminishes as the size of the sample, n, is increased. How large the sample must be before the curve loses its asymmetry will depend on the extent of the disparity between p and q. The smaller the value of p, the larger must the value of n be in order to eliminate skew.

(iv) An increasing proportion of estimates of p derived from samples of size n are nearer the true p—i.e. the initial p in the binomial $(q+p)^n$—as n increases its value. In other words, the larger the sample, the better our chance of making an accurate estimate of the initial p, assuming we do not know it.

Our dilemma is that we can never know what the probability of an event is beforehand, and we can never be sure that we have estimated it accurately from an actual series of events. What Bernoulli discovered was that certainty can be attained only by taking an infinite number of events. As this is impossible, he tried to compute the chances of getting within known limits of the 'initial' probability. He was not very successful in this for the computational work is enormous, requiring a summation of all the terms in a binomial expansion, and his results could not be reduced to a rule of general application. He calculated that, if p is 0·6 the probability of getting an estimate of p between 0·58 and 0·62 would be 0·999 in samples containing 25,550 events, 0·9999 in samples of 31,258 events, and 0·99999 in 36,966 events. Nevertheless, this approach was sensible and provided a sound basis for building a theory of sampling. Bernoulli realized that this development had value in practical problems as well as in the solution of games of chance. What was now required was a quick, simple, and generally applicable method of measuring the reliability of figures obtained from samples of observed events.

Chapter VI

THE NORMAL CURVE

THE man who solved this problem of simplifying and standardizing the work of calculating probabilities was Abraham de Moivre (1667–1754), a brilliant mathematician who, although born in France, lived most of his long life in England. At the age of 30, he was elected a Fellow of the Royal Society. Apart from his work on the theory of probability, he made important contributions to other aspects of mathematics.

De Moivre had men of wealth bring to him, as they had done in the past to Pascal and Fermat, their questions regarding chances in games of dice, roulette and cards. Thus, he acquired a wide knowledge of probability problems and in 1718 he published a book entitled *The Doctrine of Chances, or a Method of Calculating the Probabilities of Events in Play*, which he dedicated to Sir Isaac Newton. In 1733, de Moivre published privately a brief paper in Latin with the title *Approximatio ad Summan Terminorum Binomii $a+b^n$ in Seriem Expansi*, in which he presented the formula for what is now known as the Normal curve; a translation was incorporated in the second edition of *The Doctrine of Chances*, published in 1738.

In these memoirs de Moivre stated very clearly the problem we have been considering of how to handle the binomial theorem when the value of *n* is very high. His remarks are worth quoting: 'Although the solution of problems of chance often require that several terms of the binomial $(a+b)^n$ be added together nevertheless in very high powers the thing appears so laborious, and of so great difficulty that few people have undertaken the task; for besides James and Nicolas Bernoulli, two great mathematicians, I know of nobody that has attempted it; in which, though they have shewn very great skill, and have the praise which is due to their industry, yet somethings were farther required; for what they have done is not so much an approximation as the determining of very wide limits, within which they demonstrated that the sum of the terms was contained. Now the method which they followed has been briefly described in my Miscellanea Analytica which the reader may consult if he pleases, unless they rather chuse, which perhaps would be the best, to consult what they themselves have writ upon that subject: for my part, what made me apply myself to the inquiry was not out of opinion that I should excell others, (in which however I might have been forgiven), but what I did was in compliance to the desire of a very worthy gentleman and good mathematician, who encouraged me to it.'

The formula which de Moivre worked out for describing the binomial expansion when n is very large is the foundation of modern sampling methods. The graph of this formula is generally known as the Normal curve.

In order to understand the significance of this formula it will be necessary to examine some new frequency distributions. Not all frequency distributions in statistics are of the shapes shown in Charts 2, 3, 4 and 5, although these shapes are commonly found and always apply to distributions in games of chance; but all frequency distributions can be plotted on graphs in that way.

We saw in the last chapter that the average $\frac{r}{n}$ of a binomial distribution is always equal to p. It was further demonstrated that, as the size of n is increased, there is an increasing proportion of the results whose values of $\frac{r}{n}$ fall within given limits, $p \pm e$, where $e = $ a particular proportion such as 0·1 (thus 0·5 ± 0·1 which represents the limits from 0·4 to 0·6). On graphs these frequency distributions become narrower with larger samples and more and more of the total area falls between the given limits. Stated another way, the variation about the average becomes smaller as the sample size is increased.

It is always useful in statistical work to measure not merely the average but also the amount of variation in a frequency distribution. As we have seen already (in Chart 5) an identical average may arise from figures which vary considerably. Of course, an average implies variability among the values of the items in a distribution; it summarizes the central tendency of the distribution. But it tells us nothing about other characteristics. For those we need other summaries. The spread of the items in a frequency distribution from their average can also be measured and summarized. Measures of variability are usually known as measures of dispersion.

How can dispersion be measured? One way is clearly to compare the sizes of the extreme items, thus indicating the range of values. This is useful in certain circumstances but, being based on only two items, the measure can be misleading when few of the items are near these extreme values and most are found within a small distance away from the average. For example, in Chart 5 the ranges of all three distributions are from 0 to 1·0 as we know from Tables 6, 7 and 8, but the concentration of items is different in each.

A measure of dispersion which takes account of the distances from the average of *every* item is preferable to the range. One useful way of doing this is by computing another average, this time the average of the distances of all items away from the average value (i.e. the average deviation). The method is simply to take the difference between the actual result of each event (let us, as usual, call this r) and the average result which we will denote $\bar{r}$, add all these differences together and then divide by the number of events (we can call this number C because it is here the same as the number of ways of getting r results in a sample of n). Thus, the average measure of dispersion is:

$$\frac{\sum_{1}^{c}(r-\bar{r})}{C}.$$

There is, however, a qualification. An essential consequence of an average is that the sum of the deviations from it is zero, so that this method is of no use unless we treat all negative deviations (where r is smaller than $\bar{r}$) as though they are positive. The following very simple calculation will show what is meant.

Actual number of successes	Average number	Deviation from the average	Number of events	Total deviations (3) × (4)	
				With signs	Ignoring signs
r	$\bar{r}$	$(r-\bar{r})$	C		
(1)	(2)	(3)	(4)	(5)	(6)
0	2	−2	1	−2	2
1	2	−1	4	−4	4
2	2	0	6	0	0
3	2	+1	4	+4	4
4	2	+2	1	+2	2
TOTAL			16	0	12
AVERAGE DEVIATION					0·75

So, by ignoring the signs, we obtain an average value of 0·75 (i.e. 12 divided by 16) representing the average deviation each side of the mean. As, in this example, we have a symmetrical distribution it will be noted that this is the same as taking each half separately (+6 divided by 8 and −6 divided by 8). Of this frequency distribution we can say that it has an average (or mean) of 2 and an average deviation of 0·75.

The average deviation is easy to understand and it usefully summarizes the amount of variability in a distribution. When, however, a measure of dispersion has to be incorporated in a mathematical formula, the average deviation is of no value because the signs would have to be taken into account and, as a result, this measure of dispersion would always have 0 as the answer. It could therefore play no part in the formula. For this reason the deviations are usually squared. If -2 is multiplied by itself, the answer is $+4$; similarly, if -9 is multiplied by -9, the result is $+81$. The average of these squared deviations is known as the Variance. Thus:

$$V \text{ (Variance)} = \frac{\sum_{1}^{c}(r-\bar{r})^2}{C}.$$

As this represents the average of the sum of the *squares* of every difference between the actual item and the average, the next step is to find the square root of this sum. The answer is a most important measure of dispersion and is known as the standard deviation. The symbol used to denote it is σ (pronounced sigma), which is the Greek letter for small s. So the standard deviation is:

$$\sigma = \sqrt{\frac{\sum_{1}^{c}(r-\bar{r})^2}{C}} = \sqrt{V}.$$

The following shows how this measure is computed:

Actual number of successes	Average number	Deviation from the average	Deviation squared	Number of events	Total of squared deviations
r	$\bar{r}$	$(r-\bar{r})$	$(r-\bar{r})^2$	C	$(4)\times(5)$
(1)	(2)	(3)	(4)	(5)	(6)
0	2	-2	4	1	4
1	2	-1	1	4	4
2	2	0	0	6	0
3	2	$+1$	1	4	4
4	2	$+2$	4	1	4
TOTAL				16	16
STANDARD DEVIATION					1·00

This is the same frequency distribution used for illustrating the calculation of the average deviation. Again the average is 2, but the standard deviation is 1·00. This, of course, is the square root of $16 \div 16$; it happens in this example that $\sqrt{V} = 1$.

The reader may feel that this measure is not really an improvement on the simple arithmetic average which is easier to understand. In fact, the standard deviation, when understood, is just as meaningful as an ordinary average. It is entirely a question of getting familiar with it. It is of more value in mathematical calculations and is therefore generally used instead of the average deviation or other measures of dispersion. Its meaning and value will be quite clear by the end of this chapter.

The standard deviation is also very easy to compute for a binomial distribution, because it always equals $\sqrt{pqn}$ when we are interested in measuring the variability about the average of the *numbers* of successes

(r) and $\sqrt{\dfrac{pq}{n}}$ when we are interested in the variability in the *proportions*

of successes $\left(\dfrac{r}{n}\right)$ in a distribution. For example, the distribution used

above for demonstrating how the average and the standard deviations are computed is the expansion of the binomial $(0·5 \times 0·5)^4$. As $p = 0·5$, $q = 0·5$ and $n = 4$, the standard deviation of the numbers of successes (r) is $\sqrt{0·5 \times 0·5 \times 4} = \sqrt{0·25 \times 4} = \sqrt{1} = 1$. If instead of the numbers of successes we had shown the proportions $(\dfrac{r}{n} = 0, 0·25, 0·5, 0·75, 1·0)$

the standard deviation would be $\sqrt{\dfrac{0·5 \times 0·5}{4}} = \dfrac{0·5}{2} = 0·25$. So we can

say about this frequency distribution of proportions of successes that it has an average of 0·5 and a standard deviation of 0·25.

We are now able to examine some new frequency distributions which will show the significance of the formula which de Moivre worked out. And again, in order to avoid complicated mathematics, diagrams will be used to explain what the Normal curve means and how it is derived from the binomial distribution.

It will be remembered that in Charts 2, 3, 4 and 5 we measured frequency (or probability) along the ordinate and the number (r) or

proportion $\left(\dfrac{r}{n}\right)$ of successes along the abscissa of each chart. The areas of the graphs in Charts 2 and 4 were all equal and would have been in Charts 3 and 5 if the ordinate scales had been suitably multiplied to allow for the contraction of the abscissa measurements. We are now going to take the binomial distributions given in Tables 5, 6 and 7 and alter them in two respects.

First alteration. Instead of relating the frequencies (which are the probabilities) to the number of heads, we shall regard each frequency as representing the probability of getting a particular *deviation* from the average. Thus, if the probability of getting three heads in throws of 10 coins is 0·1172, this must also be the probability of getting a result 2 below the average of 5 heads. Similarly with the other results:

Numbers of heads (r)	0	1	2	3	4	5	6	7	8	9	10
Deviation from the average	-5	-4	-3	-2	-1	0	$+1$	$+2$	$+3$	$+4$	$+5$

or

Proportions of heads	0·0	0·1	0·2	0·3	0·4	0·5	0·6	0·7	0·8	0·9	1·0
Deviation from the average	$-0·5$	$-0·4$	$-0·3$	$-0·2$	$-0·1$	0	$+0·1$	$+0·2$	$+0·3$	$+0·4$	$+0·5$

These are the deviations from the expected—as well as the actual—average (the p in $(q+p)^n$). These deviations could have been used for the abscissa scales instead of the numbers in Chart 2 and the proportions in Chart 3. Instead of thinking of the probability of getting a particular result, we think of the probability of getting a particular deviation from the average result. And as this average result is in effect our 'true' p, we are really thinking in terms of deviations from this true probability (the one we are, by taking a sample, trying to find).

Second alteration. These deviations, $(r-\bar{r})$ or $\left(r-\dfrac{\Sigma r}{n}\right)$, will now be related to the standard deviations (σ) of the frequency distributions. This will be done by dividing the deviation of an individual result from the average result by the standard deviation of the distribution, i.e. $\dfrac{(r-\bar{r})}{\sigma}$. However, to counteract the effect of reducing the width of each column in this way, the heights are lengthened by multiplying the probabilities ($^nC_r q^{n-r} p^r$) by the standard deviation. The areas of the distributions are thus kept constant by the method explained in pages 37 and 38.

These alterations for the three binomial distributions, $(q+p)^n$, where n equals 10, 24 and 100, and q and p both equal $\frac{1}{2}$, are given in Tables 9, 10 and 11. Columns 2 and 4 of those tables are plotted in the graphs in Chart 6, column 2 being measured along the abscissa scale and column 4 along the ordinate scale.

There are three noteworthy features of these graphs:

(i) the shapes of all three are very similar;

(ii) the value of the middle term, which is also the maximum value, is just under 0·40 in each graph;

(iii) all the items in each appear to fall within about three standard deviations distance from the middle term.

The second and third features suggest that the curves are in fact identical in all three graphs, the apparent differences arising solely from the different number of terms in each distribution. The first graph in Chart 6 is based on a distribution containing 11 terms all of which can be plotted. The second distribution has 25 terms but only 19 of these have an ordinate value which can be shown to four decimal places and only 15 of these are really large enough to be plotted on a graph of this size. Of the 101 terms of the binomial $(0·5 \times 0·5)^{100}$ only 37 can be expressed to four decimal places (see Table 11) and of these 18 are too small to be shown in the third graph in Chart 6.

As the samples get larger, there are more and narrower columns on the graph until, on a particular size chart, each column will, for all practical purposes, become so narrow that the tops will cease to appear as steps and will become merely a large number of points so closely packed that they are indistinguishable from a continuous line. Even if we enlarge the size of the paper and make a larger chart, we can go on for ever increasing the size of the sample—we can go on, that is to say, increasing the value of n in the expansion of a binomial, $(q+p)^n$—so a sample size will always be reached when our steps will vanish into a continuous line. When this happens, the scale along the abscissa also becomes continuous.

This idea of continuity is another aspect of infinity. Just as we can go on adding (or subtracting) numbers for ever, so we can go on making smaller and smaller fractions. There is no end to 'largeness' and no end to 'smallness'; n can go on growing and $\frac{1}{n}$ can go on getting smaller.

Numbers are not only limitless in each direction, plus or minus, they are limitless in density. A continuous line simply means one which is built from variables which can have any value, however large or small.

De Moivre's formula represents the limit which the binomial distribution approaches as the size of the sample, n, is indefinitely increased, i.e. it expresses the mathematical relationships between $^nC_r q^{n-r} p^r \times \sigma$ and $\frac{r-\bar{r}}{\sigma}$ when n is infinitely large. This formula gives a continuous curve which approximates any binomial distribution when the deviations from the average result are expressed in terms of the standard deviation of the distribution. Being a continuous curve the summation of the probabilities has to be performed by a different method.

With the binomial, the probability of being within a given distance from the average is found by adding together the probabilities of getting many results, a task which, with large values of n, is really frightening. The Normal curve removes all that work. As it is a continuous curve, the area between any two vertical lines underneath it has to be computed by the methods of integral calculus which, in Abraham de Moivre's time, was a new mathematical technique. Indeed, it was the invention of the calculus which made possible this approximation of the binomial distribution by a standard formula.

As the Normal curve is an approximation to all expansions of the binomial, $(q+p)^n$, the values of the ordinates, y, and of the proportions of the area lying under any part of the curve can be computed and tabulated for general use just as factorials and logarithms have been. De Moivre calculated the area between the middle term (the average value) and the ordinates at distances equal to 1, $1\frac{1}{2}$, 2 and 3 standard deviations away from it. The name 'standard deviation' was not, in fact, used by him (it was first introduced by Pearson) but his calculation of 0·682688 is only one point out in the sixth decimal place, which is very good considering the absence of calculating equipment at that time. De Moivre also calculated the value of the deviation from the mean which divides the area of the curve each side of the mean into halves—i.e. gives limits in which one has a 50–50 chance of getting the true p. This was subsequently (in 1815) called the Probable Error and was a popular measure for a time. More detailed calculations of the values of the ordinates and of the areas were not made until 66 years after de Moivre's statement of the formula of the Normal curve, and then not in terms of the standard deviation. Other tables followed but it was not until the standard deviation was found to be so convenient a measure and came into general use that detailed tables of the curve were computed in terms of that measure by W. F. Sheppard. These tables were issued in 1902. Sheppard's tables are now in general use and an abridged version is given in Table 12.

The Normal curve is shown in Chart 7. Let us briefly note its features:

(i) It is symmetrical.
(ii) The curve is similar in shape and measurements to those in Chart 6.
(iii) The value of the middle term is 0·39894.

E

(iv) Of the total area under the curve 0·9973 falls within three standard deviations distance from the middle term. (The total area is unity because this is the sum of all the binomial terms of all possible answers).

(v) The curve is open at both ends.

The first feature was to be expected from the behaviour of the distributions in Chart 5, for the Normal curve is an approximation of the binomial when n is infinitely large. For this reason, the curve is open at both ends; it never touches the abscissa but goes on extending for ever. This is not surprising. As we saw in the previous chapter we can never be *sure* that we shall not get an extreme, freak, value. It is not impossible; there is *some* chance of getting it.

With the aid of the Normal curve the probabilities of getting a particular result can be easily obtained for all sizes of n. All that is necessary is calculation of the standard deviation which, as we have seen, is also simple.

Let us now use this curve in order to estimate the probability of getting within defined limits of the true value of p in different size samples. For convenience, we will use the figures in Tables 9, 10 and 11 because these give probabilities of getting the true p and of deviations from it calculated from expansions of binomials. We must remember, however, that these samples are all very small. For example, the probability of getting a sample of 10 containing from 3 up to 7 heads (i.e. answers within ± 2 from the average of 5) is 0·8906 (see Table 9). Instead of expanding the binomial $(\frac{1}{2} + \frac{1}{2})^{10}$ and summing these terms we could have used the values of the areas under the Normal curve given in Table 12. The standard deviation of this distribution is 1·58114, so the difference from the average of 2 is equal to 1·265 standard deviations (2 divided by 1·58114). So we look up 1·30 (which is the nearest value in Table 12) and we find that 0·4032 of the total area lies between the middle term (zero) and a point on the abscissa 1·30 standard deviations away from it. Therefore, the proportion of all samples of 10 items having from 3 to 7 heads is approximately twice this, i.e. 0·8064. This is about 10% too low but a sample of 10 is, of course, very tiny.

Let us try again, this time with a sample of 100 coins and a closer margin of error (say 5). From Table 11 we find that the sum of the probabilities of getting results of 50 heads plus or minus 5 heads is 0·7286. So the chance of getting a sample of 100 coins with an estimate of p ranging from 0·45 to 0·55 is 0·7286. Now we will get an approximation from the Normal curve. The standard deviation of the binomial distribution $(\frac{1}{2}+\frac{1}{2})^{100}$ is 5·0. Our permitted margin of error is thus exactly equal to one standard deviation. In Table 12 we find that 0·34134 of the area under the Normal curve lies between the middle and a point one deviation away; therefore, twice this, 0·68268, gives the probability of samples of 100 coins containing proportions of heads ranging from 0·45 to 0·55. This approximation is nearer the actual addition of the binomial terms than with the sample of ten. As the size of sample is further increased, the Normal curve approximation becomes even closer. As it is with large values of n that the computational labour and difficulties are greatest, the Normal curve provides that quick, simple and generally applicable method of testing the reliability of sample results which is useful. Since fairly large samples are necessary in order to get good estimates of true p, the Normal curve is most useful where it is most needed.

CHAPTER VII

THE LIMITS OF CONFIDENCE

LET us now recapitulate. The reader will remember that the probability of getting a particular attribute in a set of n events is obtained by assuming, to begin with, that certain basic things are equally likely (i.e. equally probable). One probability is derived from another probability, usiug the addition and multiplication principles. And the first probability is merely a guess.

These ideas are embodied in the binomial theorem which summarizes the number of ways in which p and q are found in n events. Expansion of the binomial, $(q+p)^n$, gives the probabilities of getting all possible sets of n, $^nC_r q^{n-r} p^r$, through all values of r from nought to n, i.e. $^nC_0 q^n$ to $^nC_n p^n$. There are thus many possible values of $\dfrac{r}{n}$ but the average of them all, added together, exactly equals p in the binomial $(q+p)^n$. The value of $\dfrac{r}{n} = p$ is also the most frequent result. For that reason it is the most probable probability.

The process of creating one probability from another is, apparently, endless. To be meaningful, our definition of probability, as a relative frequency, must imply that in a run of events the attribute will, in fact, appear that proportion of times. Indeed, the odds in a gambling game are based on this concept. Yet, as we saw in Chapter V, however long the run of events, however large the value of n, we can never be sure that the attribute will appear a particular $\dfrac{r}{n}$ times or that $\dfrac{r}{n}$ will equal p. There is no certainty about it at all. In fact, since p is the average of all possible values of $\dfrac{r}{n}$, deviations from this 'true' probability are implicit in the theorem. What we have found is that there is a growing proportion of samples which have a value of $\dfrac{r}{n}$ within a stated margin each side of true p. The probability of being *near* to p gets better as n gets bigger.

Calculating probabilities from a binomial when n is very large is a long and tedious task. An approximation known as the Normal curve provides a simple way of estimating the probability of getting values of $\frac{r}{n}$ near the true value of p. How closely the Normal curve fits the binomial distribution depends on the values of q and p and on the size of the sample, n, but the approximation is very good whenever n is very large, even though either p or q is small.

In the Normal curve, probabilities are represented by areas which have a fixed relationship with a measure of dispersion known as the standard deviation. For a given value of p, the standard deviation gets smaller as the size of the sample, n, increases. The way in which the values of $\frac{r}{n}$ cluster round the value of p can be readily seen by converting the margin each side of p into standard deviations. If this margin is denoted by e, the proportion of values of $\frac{r}{n}$ within $p \pm e$ will increase since $\frac{e}{\sigma}$ will get larger as n gets larger and σ, in consequence, becomes smaller. If we take p as 0·60 and e as 0·02, we can calculate the increased chances of being within those limits, 0·58 and 0·62, as the samples grow in size. The following table shows the figures:

Sample size n	Standard deviation σ	Limit (0·02) in standard deviations $\frac{e}{\sigma}$	Probability of being in $0·60 \pm 0·02$
100	0·04900	0·410	0·318
500	0·02191	0·913	0·637
1,000	0·01550	1·290	0·803
5,000	0·00693	2·890	0·996
50,000	0·00219	9·130	1·000
500,000	0·00069	29·000	1·000

Let us now look, in a slightly different way, at this question of defining probability. We know from Table 12 that the probability of getting values of $\frac{r}{n}$ within three standard deviations both sides of p is 0·9973. This is a very good chance. It means that 99·73% of all samples of a particular size are likely to have values of $\frac{r}{n}$ within $p \pm 3\sigma$. Only 0·27% of the samples would, therefore, have values outside those limits.

In order to see how the values of $\frac{r}{n}$ get closer to p as the sample size, n, increases we will now set the limits at distances of three standard deviations each side of p. These are the limits which almost certainly cover most samples. If we toss a coin n times there is only a 0·0027 probability of obtaining an estimate of p (i.e. a value of $\frac{r}{n}$) outside these limits.

The range of values of $\frac{r}{n}$ which would be found in different size samples when $p = 0·50$ is given in Table 13. This is the kind of result we might well get in tossing a coin whose two sides have equal probabilities of appearing. In 100 tosses we are likely to get an estimate of p between 0·35 and 0·65 because, of all possible tosses of 100 coins, 99·73% will have values between those limits. In 100,000 tosses, however, we shall find, almost for certain, that we have an estimate between 0·49526 and 0·50474. With one million tosses we can expect to get an estimate of p between 0·4985 and 0·5015. If we are working only to two decimal places we should with a sample of one million get an estimate of 0·50.

The way in which this three standard deviation band narrows and gets closer to the values of p is shown in Chart 8. On this size of paper we cannot go further. We are dealing once more with an infinite series of values. We can, of course, go on taking bigger and bigger samples and as we do so we shall find that the margin of error in the estimates of p gets smaller and smaller until finally we can imagine it to disappear altogether. Thus, the band converges indefinitely and approaches nearer and nearer to p without ever reaching it. This endless sequence of samples has values of $\frac{r}{n}$ which tend to one particular value (p) as n tends to infinity. The probability, p, may therefore be defined as the limiting value of $\frac{r}{n}$ in an endless sequence of samples.

This is a definition of probability which was propounded in 1919 by Richard von Mises. He objected to the classical theory which we have been examining, based on equally likely cases, on the grounds that it was logically inconsistent and incapable of dealing with problems where equally likely cases could not be found as starting points. To Richard von Mises the relative frequency of an attribute in an unlimited repetition of events was the only sensible basis for a theory of probability. He introduced the word 'collective' to denote any series of events or processes—such as the throws of dice in the course of a game—in which probability could be calculated. The relative frequency of an attribute in a collective becomes more stable as the number of observations is increased. If, for example, the relative frequency is calculated to two decimal places it will be found that, beyond a certain number of observations, the numbers in these two decimal points do not change although more accurate calculations would show some variation in third or later places. Beyond a still larger number of observations, the third decimal point would also become stable. It is this stable, fixed value to which the relative frequencies tend as the number of events grows larger which is really the probability of getting that attribute. More exactly a collective is 'a long series of observations for which there are sufficient reasons to believe the hypothesis that the relative frequency of an attribute would tend to a fixed limit if it were indefinitely continued. This limit will be called *The probability of the attribute considered within the given collective.* (Richard von Mises in *Probability, Statistics and Truth.*)

But this definition is only valid if the relative frequencies of an attribute in all partial sequences taken from the collective also have the same fixed limiting values. Like the original collective, these partial sequences must be capable of indefinite extension. The limiting value in a collective must not be influenced by place selection.

These ideas will be seen more clearly with an example. The following are the results of my tossing a penny 200 times, shown in the order of happening.

```
T T T H H H T T T H H H T H T H H T T T
T T T T T H T T H H H H H H T T T H T H
H H H H H H T T T H H T T H H T T T H H
T H T T T T H H T H H T T T H T H T T H
T T T H T T T T T H H H H T T T H T H H
T T H H H H H H H H H H H H H H H H H T
T T H T T H H H T H H H H H T H T T H H
T T T H T H T H T H T T T H T H H T T T
T T H T H H T T H T T T H T H T T T T T
T T H H T H T H H H H H H T H T H H T
```

The attribute we want is, say, heads and we wish to know the probability of getting heads in the throws of one penny. In this collective, the relative frequency of heads is 0·505 (i.e. 101 heads out of 200 throws). This is very near the probability value we should expect from a coin because each side is thought to have an equal chance of appearing. Let us select various sequences from this collective. The following are the relative frequencies of heads in each of the 40 results obtained by selecting every fifth throw:

Beginning with the first throw	0·550
Beginning with the second throw	0·425
Beginning with the third throw	0·420
Beginning with the fourth throw	0·575
Beginning with the fifth throw	0·525

Now let us see what answers we shall get by selecting every fourth throw instead—i.e. samples of 50:

Beginning with the first throw	0·380
Beginning with the second throw	0·600
Beginning with the third throw	0·480
Beginning with the fourth throw	0·560

As the total number of throws can be indefinitely extended, so can all of these partial sequences. The results certainly suggest that the hypothesis of a limiting value of 0·5 seems very reasonable for each of these sequences of the throws of a penny. By continuing the experiment we could be more sure.

These results are characterized by disorder or 'lawlessness'. The events have occurred in a purely random manner. In the terminology of the classical theory of probability (see Chapter IV, page 24), the events are independent and mutually exclusive; one event is not influenced by any other event. In a random series, the limiting values of the relative frequencies of getting a particular attribute are independent of all place selections; and this randomness, according to von Mises, is essential for the proper determination of probability. Because of this principle of randomness, gambling systems are useless for they usually depend on a special selection of events based on fallacious reasoning about the way in which place rearrangement effects probabilities. Therefore, Richard von Mises called it the principle of the impossibility of a gambling system. For example, some people think that if tails has just occurred, then heads must be more likely to occur on the next throw. They would be in good company, for the mathematician D'Alembert (1717–83) was convinced that this was so. We can, however, test whether probability is changed in this way by seeing what happened in the 200 throws of my penny. We will take a partial sequence by selecting only those results which have been preceded by a tail. There are 98 such results and 43 are heads and 55 are tails. Thus, in this sequence, the relative frequency of heads is 0·44 (43/98), so the chance of getting heads was certainly not improved. What, however, would be the probability if tails had already occurred *three* times? Would there not then be a better chance of getting heads? Again we will take a partial sequence from the collective of 200 throws, this time of the results which are preceded by three consecutive tails. There are 28, and of these, 13 are heads and 15 tails, giving a relative frequency of 0·465 for heads (13/28). The presumption is clearly that the limiting value of these partial sequences, if continued, would really be the same as for all the others which we have taken from this collective and the same as for the collective itself. Place selection does not affect the probability of an event; the series is truly random. The differences from the true probability—which is perhaps 0·50—arise only because the collective and the selections from it are all small.

The only satisfactory definition of probability is that it is the limiting value of the relative frequency of an attribute in a limitless random series. The assumption of equally likely cases really implies equal occurrence in some long series of events. Yet, as soon as we try to find *a posteriori* the true probability—p in the binomial $(q+p)^n$—we see that a complicated nexus of probabilities rests on it. Never can we be certain of finding a relative frequency exactly equal to this true probability; but we shall get nearer and nearer to it as the series indefinitely grows.

In Richard von Mises' view, the probability of an event can never be known beforehand. This follows from the definition of probability as a limiting value. The classical theory, however, grew out of gambling and it was the physical characteristics of the objects used in games of chance which made possible some subjective estimate of probability. As the reader will see in the next chapter, when we consider more fully the meaning of probability and randomness, these physical characteristics are akin to causes of the behaviour of a coin or die or roulette wheel, so leading one to expect certain probabilities. The proof of these hypothetical probabilities can come, of course, only from experiment. Moreover, there are many events for which no initial hypothesis about probability is possible. Therefore, we are forced to make an estimate from an actual experiment, a sample of events.

The reader will not have failed to note that a dilemma remains. If the purpose of taking a sample is to find the value of the unknown true probability, then the standard deviation will also not be known. How, therefore, can it be used for testing the accuracy of the sample estimate of p? Do we, after all, really know the limits within which the true value of p is likely to fall?

The answer is that we have an estimate of the limits in the form of the standard deviation based on the estimated p. The purpose of appending limits to an estimate of p is to indicate what margin of error may be expected from such a result. Let us be clear about this. If the true probability of an event happening is 0·35 then 99·73% of all samples of 1000 items or occurrences will show estimates of that probability between the limit of three standard deviations each side of 0·35. Since the standard deviation is 0·0155 the limits are therefore 0·35±0·0465, i.e. 0·3035 and 0·3965. This means that there would be only a probability of 0·0027 of getting an estimate outside those limits. If we were unlucky enough to get one of the extremes as the estimate of p—for example 0·3035—and we knew the standard deviation, we would know that the true probability lay somewhere between 0·3035 plus or minus 0·0465, i.e. 0·2570 and 0·3500. But we do not know that standard deviation and we have to calculate one from $\dfrac{r}{n}$ the estimate of p given by the sample. If $\dfrac{r}{n}$ (estimated p) happens to be exactly equal to the true probability (true p), then estimated standard deviation will equal the true standard deviation; otherwise the two standard deviations must be different and the question is: how different? It is important to know the likely extent of these differences.

The different standard deviations calculated from different values of p and n are given in Table 14. They are shown graphically in Chart 9. The standard deviations which have to be compared are, in fact, close to one another on each curve. Although a sample of 100 throws of a coin, for example, may theoretically give a value of 0·05 heads ($\frac{5}{100}$), the odds against such a result are exceptionally heavy. Indeed, if it happened one would immediately suspect the coin! Therefore, for the purpose of ascertaining the differences which could arise in standard deviations based on sample estimates, we need take only those standard deviations falling within values of p which are 3 standard deviations each side of the 'true' p.

Let us keep to the example of the penny because the expected true probability of getting, say, heads is 0·50 and this gives the highest value to a standard deviation $\sqrt{\dfrac{0·25}{n}}$. If we took a sample of 100 throws of this penny there would be a probability of 0·9973 that the values of $\dfrac{r}{n}$ would be within 3 standard deviations of 0·50, i.e. 0·50±0·15 = between 0·35 and 0·65. The probability of getting a value of p lower than 0·35 or higher than 0·65 would be only 0·0027. If our sample gave an extreme value of 0·35 for p we should then have a standard deviation of 0·0477 (instead of 0·05 based on $p = 0.50$) and if our sample gave the other extreme of 0·65 for p we should similarly have a standard deviation of 0·0477. Thus, in taking the standard deviation based on estimated p from the sample result we should be 0·0023 out—an error of 5% too low. But this is the maximum likely error. The probability of being within two standard deviations is as high as 0·9546, so we stand a very good chance of being within 0·40 and 0·60. The standard deviation of the binomial $(0.6+0.4)^{100}$ is 0·0490 which is no more than 0·001 different from the standard deviation with p and q values of 0·50. This is a difference of only 2% too low. If, therefore, our sample gave values for p of 0·4 or 0·6 we would estimate the standard deviation as 0·049 and be able to say that the true value of p would lie between 0·253 and 0·547 if the sample gave 0·4 as the value of p; and between 0·453 and 0·747 if the sample value of p was 0·6.

However, these are the differences which we would find with fairly small samples. What would be the size of the differences if we were to deal with much larger samples? Let us now examine the standard deviations we would get in samples of 5000 throws of a penny. Since the expected probability of getting heads is 0·50, the 'true' standard deviation is 0·00707. This is the standard deviation of the binomial distribution $(0.50+0.50)^{5000}$ and 99·73% of all samples of 5000 throws are likely to yield a relative frequency for heads inside the limits of $0.50\pm3(0.00707)$, i.e. between 0·47879 and ·052121 (see Table 13). The standard deviation of a binomial distribution with $p = 0.47879$ is 0·007065, which is practically the same as the 'true' standard deviation. The difference is 0·000006, or less than 0·1%. The same is true of the standard deviation of the binomial based on an estimate of 0·52121 for p. These differences are really so tiny as to be not worth worrying about. Moreover, these are the standard deviations estimated from two extreme estimates of p and the differences between estimated and the true standard deviation would diminish as the sample values of p move closer to the true value.

Since the standard deviation based on $p = 0.5$ is at its peak value, all estimates of the standard deviation can only be the same or smaller. The error is biased. We could, if we wished, simply take the standard deviation at the maximum value of $\sqrt{\dfrac{0.25}{n}}$. But if p were much smaller than 0·5—as, for example, the probability of getting a six in the throws of a die—the maximum standard deviation would be far too large. Chart 9 shows that the value of a standard deviation falls more rapidly as the value of p gets smaller or larger than 0·5. This means also that the difference between estimated and true standard deviations will be larger with smaller values of p, with smaller probabilities of getting an attribute. The error in taking an estimate of the standard deviation based on the sample result is thus liable to be progressively greater as p moves farther away from 0·5. But the error is not biased and the standard deviation estimated from the sample may be either smaller or larger than the true standard deviation according to which side of the true value of p the estimated value falls. We ought, therefore, to spend a little time examining the size of the error in using estimated standard deviations as measures of confidence limits when the probability of getting an attribute is much smaller or much larger than 0·5.

This time we will consider the probability of drawing balls from a box which contains ten round balls of the same size, one red and the other nine white. The procedure will be to shake the box, withdraw a ball from it and note the colour. The ball will then be replaced and the process repeated n times. The physical characteristics of the balls and the random process of selection lead one to expect a probability of 0·10 of drawing red. Now we know from Chapter 5, and from Charts 4 and 5, that a binomial distribution is skewed when p and q are not equal and n is small. If, then, we take a sample of 100 drawings from our box, we must expect some skew and recognize that the Normal curve will not be a very good fit. To assign three standard deviation limits in these circumstances gives only an approximate measure of the correct range even with knowledge of the true standard deviation because the true value of p is not central. Nevertheless, these limits will be useful for the purpose of seeing what kind of error we should get by estimating the standard deviation from a sample result. The standard deviation of the binomial $(0·9+0·1)^{100}$ is 0·03 and nearly all samples—a probability of 0·9973—will be found in a range designated by six standard deviations. Thus the range of the relative frequencies of getting the red ball is of the order of 0·18—i.e. 0·03 × 6—and the true probability is somewhere near but not exactly at the middle of that range. The true p, 0·10, will not be far from the middle so the actual range may be taken as 0·01 to 0·19. If our samples gave us either of these extreme values of p we should then estimate the standard deviation as 0·0099 for $p = 0·01$ and as 0·03923 for $p = 0·19$. These are much larger errors than we had when p equalled 0·5; they are errors of -67% and $+31\%$. The chance of getting these extreme values is, however, very small.

There is, indeed, a probability of only 0·0454 of being outside the limits of two standard deviations each side of the true value of p, i.e. $0·10 \pm 2$ (0·030) = 0·04 and 0·16. If we had these as estimates of p we should calculate the standard deviation as 0·020 or as 0·037. The errors have now been reduced to -33% and $+12\%$.

The binomial distribution $(0·9 + 0·1)^n$ becomes symmetrical when n is large. The normal curve is then a good approximation of the binomial. If there were 5000 drawings from the box, the standard deviation would be 0·00424 and the limits of confidence would be $0·10 \pm 3$ (0·00424) i.e. between 0·08728 and 0·11272. There would be little chance—a probability of 0·0027—of getting an answer outside those limits. If we were unlucky enough to get either of those extremes we would have the following two estimates of the standard deviation: 0·00399 when p is the lower limit and 0·00447 when p is the upper limit. These are close to the true value of 0·00425. Although the errors are now smaller and similar, $-0·00025$ and $+0·00023$ (or $-5·9\%$ and $+5·4\%$), they are nevertheless much larger than those found at the peaks of the curves on Chart 9.

The standard deviations in Table 14 have been plotted on logarithmic paper in Chart 10. This converts the numbers to the logarithms and shows the percentage relationship or ratios of the numbers instead of the absolute differences. We see in Chart 10 that the ratios of the standard deviations are always the same whatever the size of the sample. All the curves are identical in shape. At its top each curve is fairly flat and each falls away with increasing rapidity as the standard deviations are calculated from smaller or larger values of p. This means that the percentage differences between standard deviations increase as we move down a curve in either direction starting from the peak value. But the percentage differences *decrease* as the size of the sample increases because the segment of each curve covered by 6 standard deviations— 3 each side of p—gets smaller as we move from one curve to another down the chart. This must be so because the standard deviation for a given value of p decreases as n gets larger.

In studying the kind of differences which can be expected between standard deviations estimated from sample results and the true standard deviation we have been using extreme values. The errors would be smaller for all estimates of p nearer the true value. We can say that, provided the value of p is neither very small nor very large, the estimated standard deviation closely indicates the confidence limits, and even with small values of p the estimate is good if the sample is large.

Chapter VIII

THE MEANING OF PROBABILITY AND RANDOMNESS

ALTHOUGH we have come to the conclusion that probability is the limiting value of a relative frequency in an infinite series, this definition has a narrow application. It means that probability can only be calculated *a posteriori* from a series of past events and that it is only an estimate, because it has to be obtained from finite samples with margins of error which are themselves part of a probability calculation. We *believe* that, if we take a number of samples of size n, $99 \cdot 73\%$ of them would have values of $\frac{r}{n}$ lying between $p \pm 3\sigma$. This is an *à priori* probability. It means, from what seems to be a rational definition of probability, that $0 \cdot 9973$ is itself the limiting value of the probability of getting estimates of p between those margins if we continue indefinitely taking samples of size n. We cannot escape from the endless chain of reasoning which we examined in Chapter V.

Having obtained an estimate of probability from a sequence of past events, how can we be sure that it will apply to a future sequence of similar events? The answer is that, if the sample sequence is part of a random series, it ought to have a limiting value of $\frac{r}{n}$ equal to that for the whole random series. Therefore, the limiting value of $\frac{r}{n}$ in the *next* set of n events drawn from a continuation of that series ought to be the same as for a past set. Whenever we consider probability in the context of a series of future events, we can only estimate it beforehand. The early writers on probability arrived at an *a priori* probability from the number of alternative answers which are possible. The only test of the accuracy of this kind of subjective assessment is the limit to which the proportion of 'favourable' events $\left(\frac{r}{n} \right)$ tends in a random sequence indefinitely prolonged. But a probability calculated *a posteriori* from a sample of past events becomes a probability *a priori* as soon as we say that it is the value of p in a future sequence. What seems clear is that we can only make this substitution if the series of results is random.

The definition of a random series as one in which place selection does not affect the limiting value of a relative frequency tells us whether a series *is* random but it does not tell us *how* it became random, yet there must surely be some reason why the results distribute themselves in this way. Is it really sensible to say that the results happen without cause? Is this the meaning of chance? Certainly a series in which the relative frequency of getting a particular attribute has a limiting value, p, approached with a decreasing margin of error as the number of events increases—a limit unaffected by place selection—does not seem to be 'lawless'.

A chance event is usually defined as one which occurs in an undesigned or unforeseen manner. The event has, as it were, an accidental quality about it. Of the many alternative answers which are possible, we cannot say which one of them will appear next time. In games of chance, the course of events is determined independently of any skill on the part of the players. Indeed, in a game of pure chance—for example, snakes and ladders or roulette—no skill is necessary at all. In some games, e.g. cards, there is a mixture of chance and skill; a player needs skill only after the cards have been distributed among the players by what is considered to be a random method. A player can, of course, influence the events in a game of chance by cheating, as when marked cards give clues to players about the distribution of the cards or a die is so constructed that one side will appear more often than others. Cheating means that certain players have altered the chances of winning in their favour without the knowledge of the other players. There would be no objection to a game with a weighted die if all players knew that the die was weighted and that the expected chances of the appearance of each side were in consequence not equal. The game would be fair if all players knew from past experience what were the chances of each side's appearance or if none of the players knew the chances.

In what sense is a chance event unforeseen? I take a coin, place it heads uppermost on my thumb and flick it into the air, allowing it to drop on the carpet. If, every time I perform the operation I follow exactly the same procedure, always placing the coin heads uppermost on my thumb, why do I get what appears to be a limiting value of 0·5 for p as the number of throws increases? The results of tossing a penny 200 times were given in Chapter VII, page 53. I have now tossed another penny 300 times carefully following the procedure described above. Here are the results:

```
H H T T T T H H T T H T T H T T H T T H
H T T H T H T H H T T T T H T H H T H H
H T H T H T H T T T T H H H H H T T T T
T T T H H T H T T H T H T H T T H T T H
H H T H H T H T T H T T T T T T H H T T
H T H H H H H T H H T H T H H T H T T H
T H H H T T H H H T T H T H T T H H H T
H H H H T H T H T H T H T T H H H T T T
T H H T T H T H H H H H H H H H H T T T
H T H H T H T H T H T H H H T H H T H H
H H T T T H T T H T T T H H T T H H H H
H H H H H T H T T H H T H H H T H H H T
H H H T T H H T T H T T H H H H T T T T
T H T H T T H T T T T T H T H H T H T H
H T T T T T T H H T T H H H H T T H T T
```

The relative frequency of heads in all 300 throws is 0·513. In the first 100 of these tosses it is 0·430; in the second 100, 0·610; in the third 100, it is 0·500. Thus, all the indications are that, if this series of events were extended, we would justifiably assume that the probability of getting heads would be about 0·500. If we regard these throws as an addition to the 200 made earlier we get a relative frequency for heads of 0·510.

There are two possible alternatives—heads or tails—and they both apparently have the same chance of appearing. This means that over a long run we may expect $\frac{r}{n}$ to equal 0·500 (if that is the true probability), but the result of any *one* event is uncertain. Why is this? The only variation is in the strength of the flick of my thumb which sends the coin to varying heights in the air. The distribution of the weight of the coin is such that neither one side nor the other, over a long run of tosses, has a preponderance. The physical characteristics of the coin seem to be one reason why each side has an equal chance of being uppermost when the coin lands on the carpet. Were we to disturb the normal construction of the coin and deliberately weight one side, we should find that the frequency of getting heads and tails would change. Moreover, if by dint of a considerable amount of practice, we could control precisely the strength of the thumb flick and always be able to send the coin exactly the same height into the air with exactly the same number of turns to the carpet (or without any turns so that the coin landed the same side up as it was on the thumb) we would get a different value of p nearer to 1·0 or 0·0 for getting heads. One of the popular daily newspapers once reported that someone claimed to be able to toss a coin 100 times and get heads 99 times, thus disproving the usual assumption that heads and tails are equally likely. Such a result is undoubtedly possible but if, in fact, we had it, we should conclude either that the coin was biased or that the method of tossing was suspect.

Thus, the determination of a probability value in a game of chance may be said to depend on the physical characteristics of the article used—the coin, die or roulette wheel—and on the way in which opportunities are given for each alternative result to appear in a sequence of events. A series becomes random when a deliberate attempt is made to give each alternative a chance of appearing. Does this, however, imply that each alternative must be given, or can be given, an *equal* chance of appearance? The word 'random' is often used in this sense. For example, in their book, *An Introduction to the Theory of Statistics*, Yule and Kendall say: 'We may give a formal definition of random sampling by saying that the selection of an individual from a universe is random when each member of the universe has the same chance of being chosen.' This meaning would be unacceptable to Richard von Mises. The results of throwing a biased die would be random if the frequency of an attribute had the same limiting value in any number of the results, however selected. A biased die or coin is one which we have decided is so constructed that the frequency of the desired result (for example, a six on the die) is increased somewhat over the frequencies of the other alternative answers. A die or coin which is perfectly balanced so that the probabilities of each of the sides are equal is generally considered to be unbiased. To define random as equivalent to equal probabilities is to make the word synonymous with unbiased and the opposite of biased.

It is true that the probability of getting, say, a six on a biased die can only be found *a posteriori* by experiment. Yet the reasoning behind the calculation is exactly the same as for an unbiased die. If we found from a sample of throws that the probability of getting six on a biased die is 0·25 instead of 0·1̇6 on an unbiased, perfectly balanced die, we can still assert that it is the physical characteristics of the die which 'cause' it to produce that particular frequency and that the method of throwing produces a random series. The only difference between the old 'classical' and the later frequency theory of probability is that, with perfectly balanced articles, we may make a good *a priori* estimate of the probability of an event whereas, with articles not so constructed,

F

we are unable to do so. *Both* series are random in the sense that the determination of each of the different alternative attributes is a question of chance and the probability is properly regarded as the limiting value of the relative frequency of a particular attribute. Randomness must refer to the chance element in the determination of *each* result and not to an equality of probabilities for each attribute. Because the series of results is random the probabilities estimated *a posteriori* may be used *a priori*.

This is not wholly in accordance with the views of Richard von Mises who developed the frequency theory of probability and propounded the definition of probability as a limiting value. For him there can be no probability without a collective. In his book, *Probability, Statistics and Truth*, he states: 'We have nothing to say about the chances of life and death of an individual even if we know his conditions of life and health in detail. The phrase "probability of death", when it refers to a single person, has no meaning at all for us. This is one of the most important consequences of our definition of probability. . . . The definition of probability which we shall give is only concerned with the "probability of encountering a certain attribute in a given collective".'

To identify probability solely with a relative frequency in a collective is a severe limitation of the meaning of the word and seems, indeed, to rob it of much of its significance. Probability implies uncertainty and this uncertainty must relate to the future and in particular to future single events. When we say, 'it will probably rain', we refer to some future period of time—the next half hour or so, or tomorrow—and we use the word 'probably' because we know from past experience that the alternative of no rain is also possible in England. In another country the qualification would be unnecessary because past experience would lead us to know without any doubt that the next period would be wet or dry.

Probability presupposes a number of alternatives any one of which *can* occur but we cannot know for certain in advance which alternative *will* occur since the occurrence is a matter of chance. Although we have agreed that the only satisfactory definition of probability is that it is the limiting value of the relative frequency of an attribute in a limitless random series, this ought not to imply that it has no relevance at all to single events. As we saw in Chapter V (pages 35–38), the 'true' probability of encountering an attribute in a collective of n results is only one of many relative frequencies, $\dfrac{r}{n}$, and paradoxically there is a diminishing probability of finding $\dfrac{r}{n} = p$ precisely but an increasing

probability of finding $\dfrac{r}{n}$ within fixed limits of p with larger samples.
Without making any assumptions *a priori* about the value of p in the
binomial $(q+p)^n$, we have reasoned that a frequency definition would
imply that we should expect to find a high proportion of the values of $\dfrac{r}{n}$
falling within a narrowing margin of error about the value of p as the
number of results, n, gets larger. And the margin of error will become
infinitely small when n becomes infinitely large.

If we accept probability as being a limiting value which is approached
with increasing accuracy as n grows in size, we may rationally assume
that the reason for this is that the probability measures the strength of
the forces operative in a chance selection of each alternative attribute.
We know, for example, that if we alter the balance of a die we shall
change the probabilities of getting each side. Far from being hap-
hazard a random series represents the kind of distribution of alternative
attributes which we can logically expect if probability does affect each
individual result. Moreover, in games of chance, the series is clearly one
which can be continued for ever. Any finite selection of results—any
sample—is merely part of this infinite sequence. For this reason we are
able to use the sample to get an estimate of the probable limits of the
value of the true, initial, probability and we can predict that this
probability will apply to a future sequence of that random series.

However, we are not yet clear of our difficulties. So far we have
dealt only with the objects used in games of chance—with coins, dice
and balls in an urn. With these, we always get series which can be
continued indefinitely and which are easily seen to be infinitely long.
A random sample of the throws of a die is obtained simply from the n
throws of a die (or a throw of n dice). The size of n is determined when
we stop throwing. Outside of games of chance, in what is called practical
work, we always think of a sample as being a part of some larger
number of things or events; and we take a sample in order to discover
information not previously known about the whole and to get the
information more cheaply and more quickly than if we made a study of
all items. At times we may even get more accurate information. For
example, knowledge about family income and expenditure is usually
better based on the collection of budgets from a small proportion of
all households because those in the sample can be helped to keep good

records of their income and expenditure for the required periods. The sampling error may then be much smaller than the error which would be likely from less trustworthy figures obtained from all households. It is usually better to get a small number of reliable figures than a large number of doubtful ones.

Sometimes it is not practicable or possible to study all the items in which we are interested. Assessment of the average life of some product like an electric bulb requires complete consumption of each article. To base the calculation on consuming the total output would be expensive and silly. The knowledge we want can be obtained satisfactorily from samples which comprise very small proportions of the total output of the product in a given period.

In what sense does probability arise in these practical problems? Can we sensibly talk of the probability of finding in a community a household which has a total family income of x or more pounds a year? Is there a probability of an electric bulb failing after only 100 hours of life? What do we really mean when we talk of the probability of dying?

Let us look more closely at the last question. We must, of course, be more precise in our definition of the probability of dying. Clearly a person aged 80 is more likely to die in the next ten years than someone aged 15 years. We must, therefore, ask what is the probability of a person aged x years (let us say 20) dying from natural causes before attaining the age of y years (let us take this as 50). Our first sample (n) could comprise all persons in the world who today have reached their 20th birthday. Having obtained this sample we should need to keep a record of all these people during the next 30 years, noting those (r) who die from natural causes. When the survivors reach the age of 50, which they will all do on the same day, we can then calculate the proportion, $\frac{r}{n}$, who have died. Complications would arise from the occurrence of deaths from accidental causes for we do not know whether those killed in this way would otherwise have lived on to reach the age of 50 years. So we should have to exclude all accidents from the sample. For the present we can oversimplify and assume that there are no accidental deaths. We are interested only in the meaning and significance of probability in practical problems and we need not complicate our thinking by trying to solve computational difficulties. The discussion is necessarily theoretical for the present because we would not, in practice, possess observations of all persons in the world arriving at their 20th birthday today or any other day. It is useful, however, to assume that we can do so.

Is this proportion, $\frac{r}{n}$, a good estimate of some unknown relative frequency, p, which is the true probability of dying between the ages of 20 and 50, applicable to all future persons aged 20 years? If we imagine our sample to cover all persons in the world who today have reached the age of 20 we certainly have a very large sample—perhaps 5 million persons. As we saw in Chapter VII, a value of $p = 0.5$ gives the highest estimate of a standard deviation with a given value of n (see Chart 10), so the maximum standard deviation would be $\sqrt{\dfrac{0.5}{5,000,000}}$. Thus, there would be a probability of 0.9973 of finding the true probability of dying between those ages within 0.0000095 of the estimate obtained from the sample. The limits may indeed be narrower.

If, however, this is a good estimate of the true probability of dying, what is the collective—the infinite random series—of which this sample is a part? Clearly we *can* think of this series as one which extends indefinitely into the future. And it has a long extension into the past. We can, for example, add into our sample all those who reached their 20th birthday yesterday, then the day before yesterday, the day before that, and we can imagine additions back to the 20th birthday of Adam and Eve. Is this a limitless random series in which the probability of dying, p, is the limiting value of the relative frequency of deaths as the number of persons studied, n, tends to infinity? If it is, we should expect the estimate of this probability to be unaffected by place selection. Whether we selected a sample of all those who attained their 20th birthday on June 1st, 501 B.C., or on January 23rd, A.D. 1796, or on September 3rd, 1910, should make no difference to the limiting value (p) of the relative frequency $\left(\dfrac{r}{n}\right)$ found in each selection. We happen to know, however, that the choice of sample does make a difference. The proportion of persons dying between the ages of 20 and 50 has changed considerably in the last one hundred years alone and the differences found are by no means random, for the risk of dying in many countries has been declining continuously for a long time. Moreover, even a partial sequence of persons with their 20th birthday today might well show a relative frequency of deaths quite different from the total, for the proportion of persons who die between 20 and 50 differs enormously in different parts of the world.

Thus it is clear that this sequence of observations is not a random series in which the proportion of persons dying between 20 and 50 tends to a fixed limit which we can justifiably term the true probability of death between those ages. There are apparently many different probabilities of death between those ages, probabilities which differ according to the dates when people live and according to the geographical location of their lives. We have to be very careful before we conclude that a limitless sequence of observations is a collective or that a particular partial sequence is part of only one collective.

Although a relative frequency can be taken as the calculation *a posteriori* of an unknown probability, it becomes a probability only when we are able to use it *a priori* either as a prediction about the likely course and pattern of future events or as an estimate of the relative frequency of the occurrence of an attribute in some past sequence about which we have no other information. We are able to do this because probability is really, as we have already noted, a measure of the strength of the forces operating in a chance selection of each alternative result. In this sense we have to think of the probability of dying between 20 and 50 in terms of the risks to which all persons aged 20 are exposed. If they are all subject to the same risks we may imagine that, if the number of persons exposed to these unchanging risks went on increasing, the relative frequency of the number dying before reaching the age of 50 would tend to a limit as the total number of persons became infinite. The emphasis must be on the 'exposure to risk'. We may say that the population of a country in a short space of time (say a year) is subject to forces which give rise to a particular probability of dying. In so far as all persons in the country are, broadly speaking, subject to similar circumstances—climatic conditions, medical knowledge and services, housing, sanitation, education, adequate feeding, and so on— an indefinite extension of the number of those persons subject to the same conditions would be the collective in which the limiting value of the probability of dying between 20 and 50 is p. It is, of course, not so easy to imagine an indefinite extension of the persons aged 20 exposed to these risks as it is to imagine an endless sequence of coin tosses. In tossing a coin an investigator has control of the objects and the conditions and can clearly, himself, go on with the act of tossing. In practice, there are many difficulties because, in the thirty-year period needed for the study, the conditions are unlikely to remain unchanged and the probability of dying estimated from this sample may well not be the probability which is applicable to persons aged 20 at the end of this period (i.e. when those in the sample are aged 50). In other words, it is a probability which applies only to those born round about a given date in a given locality.

Moreover, the way in which random selection occurs is not always clear. It is not easy to see, for example, how the many alternative causes of illness and of death affect the members of a community and in what way selection occurs. Can we really be sure that all the people who catch cold in January do so by some unknown process of random selection and that throughout the years everyone will have a similar chance of catching cold? Or are there causes which, if better understood, would reduce or even eliminate colds? The borderline between statistics which measure the effect of the operation of some special factor and truly random results is, at times, very blurred. The stability of the proportions found in some of the earliest statistics convinced many persons that it must be evidence of divine laws. It was even thought by men of ability and influence like T. R. Malthus (1766–1834) and J. B. Say (1767–1832) that a decrease in the proportion dying from one cause would be compensated by an increase in the proportion dying from some other. Indeed, the idea that births, deaths, marriages, suicides and accidents occur inside a community with a stability which suggested divine laws also threw doubt on whether human free will existed. Were not these stable proportions evidence that most events were determined beforehand? One of the most brilliant and imaginative statisticians of the early nineteenth century, L. A. Quetelet (1796–1874), was foremost in propounding the view that physical laws determined the behaviour of people and that human will had no part in producing these results.

The truth is that there is far less precision about the meaning and measurement of probability in economic, social and other phenomena than in games of chance. Yet, in collecting information by means of samples, the probability concept is quite clear. But it is important to distinguish between the underlying probability of some happening— for example, the probability of death—and the probability of getting, in a sample, information about a larger finite number of items. Thus we can use a smaller sample to estimate probability simply because it would be expensive and impractical to estimate the probability from a larger sample or we can use a sample to ascertain information about a total without in any way regarding the proportions found in the sample as probabilities. The use of our knowledge about probability in sampling is the subject of the next chapter.

CHAPTER IX

THE IMPORTANCE OF EQUALLY LIKELY CASES

GENERALLY speaking a sample is regarded as a small part of some larger number of observations. It is customary to refer to the total to be sampled as the 'population' or 'universe'. These words have a wide meaning and they comprise the total of any kind of thing or event. Populations (or universes) are often divided into (a) hypothetical and (b) existent, or into (i) infinite and (ii) finite. All the possible throws of a die would be considered as an example of a hypothetical population since they cannot be stored or be seen all at once. They are transient things existing in time but not in space. But the total number of electors or the total number of farms in an area would be regarded as examples of existent populations. So would all the bags of wheat in a store. Hypothetical populations are usually infinite—there is no limit to the number of throws of a die—and existent populations are usually finite.

Although this is the usual classification of populations, it is confusing because the word 'population' (or 'universe') is really used in two quite different senses. Sometimes it is regarded as a random series, a collective in Richard von Mises' terminology, and sometimes it is taken to mean the collection of items possessing different attributes—really the alternatives, like the sides of a die or the different coloured balls in an urn. Strictly, a random sample can only be part of a collective, which is an infinite series. The idea of drawing a sample from a finite population seems to be in conflict with all that has been written about probability in this book.

The confusion occurs because, in many practical problems, the number of items possessing the attributes to be studied is very much larger than the size of the sample necessary to give the information we require accurately and quickly. Indeed, this is usually the *raison d'être* of sampling. If we need to know how many people in the United Kingdom wear spectacles, there is little point in surveying 50 million persons if the answer can be got from a sample of 1000. The cost of sampling would be so very much less. Yet, there is really no difference between the concept of the six sides of a die and the millions of people who comprise the population of a country. Each represents the total number of alternative items from which we can *create* probabilities. It is this

70

creation of probabilities which is important in sampling. We ought not to say that the probability of getting a six on a die is 0·16 because it is one of six sides but because this is the limiting value of the relative frequency of getting a six in an indefinite number of chance selections of each side. Similarly, we must not say that the probability of finding persons who wear spectacles is $p = \dfrac{r}{n}$ because that is the proportion who wore spectacles on a recent date in the whole population of the United Kingdom (n), but because that is the limiting value of $\dfrac{r}{n}$ in an infinite sequence of chance selections of every member of the population on that date. Moreover, we ought not to say that this relative frequency is necessarily the same thing as the probability of a person having to wear spectacles in the United Kingdom. It is solely the probability *in the collective which is created by the random selection of members of the population.* It may, or it may not also be a probability in the other sense.

We always have to be careful in talking about drawing samples from a population or a universe. If we use the word 'population' to mean the total number of alternative results we must not make it mean also the number of items appearing in a particular random series. Therefore, in this book, the word 'population' refers only to the parental aggregate of attributes. The word 'collective' is used to denote the random series from which a sample may be drawn.

This distinction between a population and a collective is most important. We take a sample in order to get information about a population. In doing so we convert the information about the population into probabilities and form a collective. The sample is not part of the population although it seems to be so whenever, as often happens in practice, it is much smaller. In deciding to collect information by means of a sample we have, therefore, to determine whether the total about which we require information is itself part of a collective or whether it is a population from which a collective has to be created by random selection from the items in it. As we saw in Chapter VIII, when trying to define the probability of death between the ages of 20 and 50, it is not always easy to be sure, outside of games of chance, that we are dealing with a collective. It is useful to know when we are, because a collective is a random series of results, so any part of it, however selected, is automatically a random sample. On the other hand our treatment of a population has to be quite different; we have to *create* a collective by making a chance selection from all the items in the population. The sample which we make is, in effect, a segment of an infinite random series, a segment of a collective. From our knowledge of probability theory we can decide in advance what size of sample will give an estimate, within a stated margin of error, of the proportion of items in the population having the special attributes. We can do this with the aid of the formula for the Variance (see pages 42, 43):

$$\sigma^2 = \frac{pq}{n}.$$

If we multiply both sides by n and then divide by σ^2 we get:

$$n = \frac{pq}{\sigma^2}.$$

Since we do not know the value of p we can get some idea of its order of magnitude by taking a small random test sample of, say, 100. Let us imagine that we have done this and have obtained a relative frequency of 0·3 for the attribute about which we require information. If, therefore, we wish to get an estimate within 10% of the true proportion, the margin of error must not be more than about 0·03 each side of p. How sure of this do we want to be? There is a probability of 0·9973 of getting a relative frequency, $\frac{r}{n}$, which is within 3 standard deviations of the true proportion and a probability of 0·9546 of getting it within 2 standard deviations. Let us decide to accept as satisfactory a margin based on 2 standard deviations. This means that we want a sample size which will have a standard deviation of 0·03 ÷ 2, i.e. 0·015. We now have all the figures we need to calculate it:

$$n = \frac{(0\cdot3)(0\cdot7)}{(0\cdot015)^2}$$

$$= \frac{0\cdot21}{0\cdot000225}$$

$$= 933.$$

In round figures, we can say that a sample of 1000 will provide an estimate of p within the stipulated limits of confidence.*

* Really the sampling problem is exactly the same as in a game of chance. There is no fundamental difference at all between a sample created by the throws of a die and a sample created by a chance selection from all the items in a population such as the total number of voters in a country or the total number of retail shops. The

Now the proportion of those in a population having a particular attribute has meaning in arithmetic only when each individual in the total is given an equal value. This equality is fundamental to rational quantitative thinking. To be able to say that 4 is twice the value of 2 requires that we give each of the units in 4 $(1+1+1+1)$ exactly the same values. This is so in statistics also. Whenever we count persons or things we invest each with an abstract arithmetical equality. For this reason we are able to state not merely that 26 is one-half of 52 but that the 26 red cards in a pack of 52 playing cards represent $\frac{26}{52}$ (or $\frac{1}{2}$) of the total. Since the remaining 26 cards are black we can assert that the number of red cards in the pack is *equal* to the number of black cards.

In order to reflect accurately in a sample the fraction of a population possessing particular attributes, the probabilities of getting every one of the alternative members of the population must be equal. The probability of getting these attributes then becomes equal to the sum of the number of equally likely 'favourable' cases expressed as a proportion of the total number of possible equally likely cases. And this is the same as the proportion of the number of individual items in the population with those attributes. Thus, we have gone full circle back to classical ideas. Although we may justifiably object to defining probability in terms of equally likely cases because this is tautological and we cannot know that the cases really are equally probable, nevertheless, in sampling, we have to go out of our way to make sure that the probabilities of getting each of the items in the population are equal in order to obtain a true answer. If there is a tendency to select particular units in a population more frequently than others, the limit of the proportion (the relative frequency) of those items in the collective must be greater than their proportion in the population. If, for example, a six-sided die is so constructed that the probability of getting a six is 0·25, then it is impossible from a sample of throws of that die to estimate accurately the proportion of the sides which are marked with a six. Our estimated proportion would be within a probability margin of error about 0·25 instead of being about 0·16.

only difference is in purpose. With a sample made by a chance selection from a population like a die, the interest is in betting about the individual results or about very small clusters of results, whereas with a sample derived from a population in human affairs and the physical sciences, where the number of items in the population may be very much larger than the size of the sample, the interest is usually in obtaining cheaply and quickly a reliable estimate about the information to be found in the population. We would not create a sample in order to find out what proportion of all persons who are eligible to vote would, on a certain date in the past, have voted for the different political parties, for we have that information already with sufficient accuracy.

The reader will recall that, in the previous chapter, we decided that the determination of a probability value in a game of chance depends on the physical characteristics of the article used—the coin, die or roulette wheel—and on the way in which opportunities are given for each alternative result to appear. We can get *equal* probabilities by throwing a perfectly balanced coin or die, or by making a series of chance selections from a pack of unmarked playing cards, all of which look identical, replacing each card drawn and shuffling the pack each time so as to be sure that every card can be selected only by chance. These equal probabilities in a game of chance imply that the articles used and the method of selection are unbiased. The results form one kind of random series; and we know that other random series can be obtained from alternatives which have unequal probabilities. In practical problems, when a sample is being drawn for the purpose of finding information about a larger population, the creation of equally likely cases is essential if the limits of the relative frequencies of the attributes in the collective are to be identical with the proportions of those same attributes in the population. It is for this reason that the words 'random sampling' are usually made synonymous with the drawing of a sample in which every member of the population is given an equal chance of inclusion.

In practice, the selection of a random sample is by no means so easy as in a game of chance. If we want to ascertain how many people in the United Kingdom own certain appliances, we cannot write the names of each person in the country on a card, shuffle the millions of cards and make a chance selection of, say, 2000. And, if we could, the task of giving each member an equal chance of being included in the sample would be immense. Less cumbersome and less costly methods have to be devised. These vary according to the character of the population and the kind of information required as well as the difficulty of making a random selection. Sample design is a study by itself. However, one theoretical feature must be noted. Replacement of each item drawn from a population is imperative if the limiting value of the relative frequency of an attribute in the collective is to be identical with the relative frequency in the population. With the throws of a coin or a die, replacement is automatic because the alternative attributes are an integral part of the object being thrown. When the members of a population are physically separate, the withdrawal of one member without replacement alters the proportion of the remainder. The probability of getting each successive result is then determined by the results of previous drawings.

The frequency distribution of all possible samples of size n drawn without replacement from a population of size N is known as the hypergeometric. This differs from the binomial in that conditional probabilities cannot be expressed by a stable p and q, as in $p^r q^{n-r}$, although the number of ways of getting combinations of r 'successes' in n events is given by the same formula, $\dfrac{n!}{(n-r)!\,r!} = {}^nC_r$, as in the binomial. Thus, we have to exchange the expression, $p^r q^{n-r}$, for another which takes account of the changes in probabilities on each drawing. Every member of the population is deliberately given an equal chance of inclusion in a sample drawn without replacement, for only in this way can the proportion in the population be accurately reflected in the sample, but the ratio of attributes remaining in the population is changed after each drawing.

The generalized formula for the hypergeometric sampling distribution is

$$ {}^NC_r \left[\frac{R}{N} \times \dots \times \frac{R-r+1}{N-r+1} \right] \times \left[\frac{N-R}{N-r} \times \dots \times \frac{(N-R)-(n-r+1)}{N-n+1} \right]. $$

This formula can be written in terms of factorial numbers thus

$$ \frac{n!\,R!\,(N-R)!\,(N-n)!}{r!\,(n-r)!\,(R-r)!\,[(N-R)-(n-r)]!\,N!} $$

(See Appendix, p. 79.)

From this formula the probability of getting each value of $\dfrac{r}{n}$ can be calculated for all values of r from $r = 0$ to $r = n$ in samples of size n drawn without replacement from a population of size N. When sampling is done with replacement, there is a probability distribution—and only one distribution—for each size of sample, with no limit, for it can be larger than the population. When sampling is done without replacement, a similar infinite sequence of results for a *given* size of population is impossible. There is a different probability distribution for every different proportion of the population, $\dfrac{n}{N}$, found in a sample for all sample sizes from $n = 0$ to $n = N$. A sample cannot be larger than the population. As it approaches the size of the population, the limits of confidence must decrease faster than for samples of the same size obtained with replacement; and when $n = N$ there can be no error at all.

However, the size of a sample drawn without replacement can be increased indefinitely for a constant fraction of the population, $\dfrac{n}{N}$, by imagining a simultaneous increase in the population size. For example, we can have samples of 5 cards drawn from a pack of 52 cards; samples of 10 cards from 2 packs; samples of 15 from 3 packs; and so on for ever. When dealing with games of chance (Chapters IV and V), we saw that a sample can be regarded as n throws of one die or as one throw of n dice. Since we are interested in ascertaining the effect of sample size on the limits of confidence, the precise content of a population is not important, provided the relative frequency of a 'success', $\dfrac{R}{N}$, is the same in all sizes of the population, so that the probability is always the same on each first drawing. What, therefore, we get with the hypergeometric formula is not only a probability distribution for every possible size of sample, without limit, but one also for all possible fractions of the population contained in the sample. Thus, there is a collective, like that shown in Chart 8, for every separate value of $\dfrac{n}{N}$.

We have now to see what is the shape of the hypergeometric distribution. Both the binomial and the hypergeometric contain the factor ^nC_r and, whenever the value of $\dfrac{n}{N}$ is small, the conditional probability $\left[\dfrac{R}{N} \times \ldots \times \dfrac{N-r+1}{R-r+1}\right]$ cannot be very different from the binomial probability $p^r q^{n-r}$. For example, the probability of getting a combination of 2 picture cards in a sample of 5 cards drawn without replacement from a population of 52 $\left(\text{i.e. } \dfrac{n}{N} = \tfrac{5}{52} = 9{\cdot}6\%\right)$ is 0·025; with replacement, the probability is $(0{\cdot}231)^2 (0{\cdot}769)^3 = 0{\cdot}024$. (See Appendix, p. 79.) It is clear that the shape of the hypergeometric curve must be similar to the binomial for low values of $\dfrac{n}{N}$, being identical when $n = 1$ (for there can be no replacement then), and narrower for higher values. The dispersion of the hypergeometric distribution gets smaller and smaller as larger proportions of the population are transferred to the sample. Like the binomial, this distribution is asymmetrical for small values of n and N, when p does not equal q, but it becomes symmetrical as n and N get larger, i.e. for increases in n with $\dfrac{n}{N}$ held constant. The average relative frequency of a success, $\dfrac{r}{n}$, in all possible samples of one size is the same in both types of distribution and is equal to the proportion $\left(p = \dfrac{R}{N} \right)$ in the population. The true answer is also the most probable.

The standard deviation of all the possible estimates of the number of 'successes' (r) in the hypergeometric distribution is $\sqrt{npq} \times \sqrt{\dfrac{N-n}{N-1}}$ and of the proportion of 'successes' $\left(\dfrac{r}{n} \right)$ it is $\sqrt{\dfrac{pq}{n}} \times \sqrt{\dfrac{N-n}{N-1}}$. Since the standard deviation of the binomial distribution is $\sqrt{npq}$ for r and $\sqrt{\dfrac{pq}{n}}$ for $\dfrac{r}{n}$, the factor $\sqrt{\dfrac{N-n}{N-1}}$ indicates the extent to which the standard deviation is smaller in samples drawn without replacement than in the same size samples drawn with replacement. This multiplier $\sqrt{\dfrac{N-n}{N-1}}$ is usually known as the finite population correction.

If the deviations in sample estimates of the value of $\dfrac{R}{N}$ are expressed in terms of the standard deviation, $\sqrt{\dfrac{pq}{n}} \times \sqrt{\dfrac{N-n}{N-1}}$, the hypergeometric curve can be approximated by the Normal curve. This approximation is good even for small samples when $p = q$ and, although less satisfactory when p and q differ, the approximation improves with larger samples for a constant $\dfrac{R}{N}$. The Normal curve can, therefore, be used to calculate the limits of confidence for estimates of p in samples drawn without replacement.

Let us imagine that we have a box containing 300 red and 700 white balls. This is a finite population of 1000 members, from which we will draw, with replacement, a random sample of 500. The estimate of the proportion of red (or white) balls given by this sample will be subject to a margin of error measured by the standard deviation $\sqrt{\dfrac{0\cdot3 \times 0\cdot7}{500}} =$ 0·0205. This means that there is a probability of 0·68268 that samples of 500 drawn from this population will give estimates of the proportion of red balls in the box within $0\cdot30 \pm 0\cdot0205$; and a probability of 0·9545 (two standard deviations) that the samples will give estimates between 0·259 and 0·341, i.e. $0\cdot3 \pm 0\cdot041$.

If, from the same population, we now draw a random sample of 500 without replacing each ball after it has been selected, there will be 500 balls out of the box and 500 remaining inside. The proportion of red balls in the sample of 500 different balls taken out of the box will lie between limits of confidence calculated by reference to the standard deviation of the hypergeometric distribution. This is $0.0205 \times$ $\sqrt{\dfrac{1000-500}{1000-1}} = 0.0205 \times 0.7075 = 0.0145$. Using the Normal curve again, this means that of all the possible samples of 500 taken without replacement from this population of 300 red and 700 white balls, a proportion of 0.68268 will show a relative frequency of red balls between 0.30 ± 0.0145 (i.e. between 0.2855 and 0.3145); and 0.9545 will show relative frequencies between 0.30 ± 0.0290 (i.e. from 0.271 to 0.329).

The largest possible sample that can be drawn without replacement from this population is 1000, when all the balls originally inside the box will now be outside it, and the sample will be identical with the population, so the proportion of red balls in the sample must be exactly right.

The standard deviation is $\sqrt{\dfrac{0.3 \times 0.7}{1000}} \times \sqrt{\dfrac{1000-1000}{1000-1}} = 0.0145 \times 0 = 0$.

Thus there is no sampling error, as we would expect. With replacement, however, drawings can be indefinitely made and a sample of 1000 balls randomly selected from this population would have a probability of 0.68268 of giving a relative frequency of red balls within a margin of error of 0.0145 each side of the true figure of 0.30; and a probability of 0.9545 of being within 0.30 ± 0.029. This error is the same as would be obtained with samples of half this size taken without replacement.

So big a fraction of the population in a sample is unusual. In most practical work, even a large sample is likely to represent only a small percentage of the population (which, in economic and social statistics, may contain millions), and the values of the hypergeometric are almost the same as those of the binomial distribution. Wherever a sample is taken for the purpose of getting information cheaply and quickly about a much larger population, the sampling procedure often requires the random selection of n different members of that population. A sample then appears to be part of the population although it is really a separate random series which is part of a collective. The binomial standard deviation is often used instead of the standard deviation of the hypergeometric distribution in the knowledge that it will overstate very slightly a margin of error which is, itself, an estimate (because the true value of p is not known). What is essential in sampling, with and without replacement, is that every member of the population is given an equal chance of being selected.

APPENDIX

Sampling without Replacement

This can be most easily demonstrated with an example. If we thoroughly shuffle a pack of 52 playing cards and deal five cards from it, what are the chances of getting two pictures? The ten different ways in which this can happen $\left({}^5C_2 = \dfrac{5!}{3!\,2!} = \dfrac{120}{6 \times 2} = 10\right)$ are set out below, denoting a picture by the letter P and any other card by the letter Q:

First drawing	Second	Third	Fourth	Fifth
P	P	Q	Q	Q
Q	P	P	Q	Q
Q	Q	P	P	Q
Q	Q	Q	P	P
P	Q	P	Q	Q
Q	P	Q	P	Q
Q	Q	P	Q	P
P	Q	Q	P	Q
Q	P	Q	Q	P
P	Q	Q	Q	P

Now the *initial* probability of getting a picture is the proportion of the number of picture cards (12) to the total number of cards in the population (52), i.e. $\frac{12}{52} = 0 \cdot 231$; and the initial probability of not getting a picture is therefore $\frac{40}{52} = 0 \cdot 769$. These are the values of p and q in the binomial for samples of 5 cards drawn with replacement from a population of 52. They are, however, only the probabilities on the first drawing of five cards without replacement. Thereafter, the probabilities at each draw must change according to the results of previous deals. If we get a picture on the first drawing, the proportion of pictures in the remaining 51 cards is now $\frac{11}{51}$ ($= 0 \cdot 216$) so the probability of getting a picture on the second draw is $0 \cdot 216$, and the probability of not getting a picture on the second draw is $\frac{40}{51}$ ($= 0 \cdot 784$). If a picture is obtained for the second time on the second draw, this would leave 10 pictures in the remaining 50 cards, giving a probability of $0 \cdot 20$ for getting a picture on the third draw, and a probability of $0 \cdot 80$ ($\frac{40}{50}$) for not getting a picture on that draw. A card without a picture on the third draw

G

means that the probability of getting a picture on the fourth draw becomes $\frac{10}{49}$ (0·204), and of not doing so, $\frac{39}{49}$ (0·796). And so on. The probability of getting two pictures in five cards drawn without replacement is calculated by multiplying the conditional probabilities for getting results in a particular order. Here is the calculation for the first row of the ten combinations shown on page 79:

First	Second	Third	Fourth	Fifth
P	P	Q	Q	Q
$\frac{12}{52}$	$\frac{11}{51}$	$\frac{40}{50}$	$\frac{39}{49}$	$\frac{38}{48}$

$$0·231 \times 0·216 \times 0·800 \times 0·796 \times 0·792 = 0·025$$

This is the probability of getting two pictures in five cards in the order PPQQQ. Let us now see what are the probabilities of getting the same combination in different orders:

$$QPPQQ = \frac{40}{52} \times \frac{12}{51} \times \frac{11}{50} \times \frac{39}{49} \times \frac{38}{48}$$
$$0·769 \times 0·235 \times 0·220 \times 0·796 \times 0·792 = 0·025$$

$$QQPPQ = \frac{40}{52} \times \frac{39}{51} \times \frac{12}{50} \times \frac{11}{49} \times \frac{38}{48}$$
$$0·769 \times 0·765 \times 0·240 \times 0·224 \times 0·792 = 0·025$$

$$QQQPP = \frac{40}{52} \times \frac{39}{51} \times \frac{38}{50} \times \frac{12}{49} \times \frac{11}{48}$$
$$0·769 \times 0·765 \times 0·760 \times 0·245 \times 0·229 = 0·025$$

$$PQPQQ = \frac{12}{52} \times \frac{40}{51} \times \frac{11}{50} \times \frac{39}{49} \times \frac{38}{48}$$
$$0·231 \times 0·784 \times 0·220 \times 0·796 \times 0·792 = 0·025$$

Clearly we need not go on. Whatever the order of the results, multiplication of the conditional probabilities gives the same overall probability of getting a sample of 5 containing 2 picture and 3 non-picture cards. The reason for this is obvious. The numerators always contain the same figures, 12 and 11 (for the two picture cards) and 40, 39 and 38 (for the three non-picture cards), and the denominators always contain the figures, 52, 51, 50, 49 and 48. Since the result of multiplication is not affected by the order of the figures being multiplied ($2 \times 4 = 4 \times 2$), the probabilities must be the same for each order.

This provides a useful basis for generalization. Let R represent the members of the population, N, which possess a certain attribute, so that $N-R$ indicates the number of those who do not possess the attribute. In a sample of size n we can assume that the first r consecutive drawings are all 'successes', and the following $n-r$ drawings are 'failures'. The probability of a 'success' on the first random selection is $\frac{R}{N}$ (equivalent to p in the binomial); on the second it is $\frac{R-1}{N-1}$; on the third, $\frac{R-2}{N-2}$; and on the rth selection it is $\frac{R-r+1}{N-r+1}$. On the next random drawing, the probability of getting a 'failure' is $\dfrac{N-R}{N-r}$, then $\dfrac{N-R-1}{N-r-1}$,

then $\dfrac{N-R-2}{N-r-2}$, and so on for $n-r$ consecutive drawings until the nth

and last selection is reached when the probability is $\dfrac{(N-R)-(n-r+1)}{N-n+1}$.

The generalized formula for the hypergeometric sampling distribution is thus:

$$^nC_r\left[\frac{R}{N} \times \cdots \times \frac{R-r+1}{N-r+1}\right] \times \left[\frac{N-R}{N-r} \times \cdots \times \frac{(N-R)-(n-r+1)}{N-n+1}\right].$$

This describes the probability of getting relative frequencies of r in samples of size n, randomly selected without replacement from a population of size N. As with the binomial, the number of different

ways of getting identical values of $\dfrac{r}{n}$ is ^nC_r; but the rest of the formula

takes the place of $p^r q^{n-r}$ in the binomial.

The formula can be written more conveniently in terms of factorial numbers. It will be recalled that a number, n, multiplied by its preceding numbers back to $n-r+1$ (i.e. r figures) can be expressed as factorial n

($n!$) divided by the unwanted tail, $(n-r)!$, thus: $\dfrac{n!}{(n-r)!}$. The numerator

of $R \times (R-1)$ back to $(R-r+1)$ is, therefore, equivalent to $\dfrac{R!}{(R-r)!}$;

and the numerator of $(N-R) \times (N-R-1)$ backwards to $(N-R)-$

$(n-r-1)$ is equivalent to $\dfrac{(N-R)!}{[(N-R)-(n-r)]!}$. The two denominators are

continuous and represent the number N multiplied by its preceding numbers down to $N-n+1$, i.e. n figures. This can, therefore, be

written as $\dfrac{N!}{(N-n)!}$. And, of course, ^nC_r is $\dfrac{r!(n-r)}{n!}$. So the formula

now becomes:

$$\frac{n!}{r!(n-r)!} \times \frac{R!}{(R-r)!} \times \frac{(N-R)!}{[(N-R)-(n-r)]!} \div \frac{N!}{(N-n)!}$$

$$= \frac{n!}{r!(n-r)!} \times \frac{R!}{(R-r)!} \times \frac{(N-R)!}{[(N-R)-(n-r)]!} \times \frac{(N-n)!}{N!}$$

$$= \frac{n!R!(N-R)!(N-n)!}{r!(n-r)!(R-r)![(N-R)-(n-r)]!N!}$$

Chapter X

MEASUREMENT BY SAMPLE

A SAMPLE in the sense of a small part of some larger quantity will give reliable information about the contents of the larger quantity if the individual items are randomly selected, so forming a collective in which the relative frequency of those items having certain attributes tends to a limiting value which is identical with the relative frequency of items with the same attributes in the population. The information obtained is necessarily about proportions only, being a classification of results of items according to whether or not they possess the required attributes. At bottom, all classification is a process of twofold separation of things by noting the presence or absence of an attribute. If we want to classify all persons in the United Kingdom we can proceed in the following way:

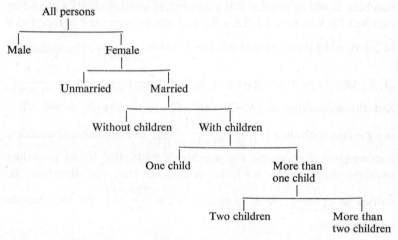

And so on, according to the kind of information wanted.

Clear and rational classification is very important in statistics because we are rarely interested in the number of items in a single class only, but in knowing also some details about the contents of the class. Take, for example, the statistics in Table 15. The total number of persons in the United Kingdom in 1960 is of very little interest by itself. It becomes interesting when compared with totals at other dates, or with the totals of other countries, or when, as in this table, it is broken down into other classes so that we can see the relationship of one figure with another and, in particular, of the number in one class with the total.

In a sample drawn from a population, each number in each class may, in turn, be regarded as a 'success' (r) which may be expressed as a proportion $\dfrac{r}{n}$ of the total and so provide an estimate of the proportion in the population (p), which is the limiting value of the relative frequency $\dfrac{r}{n}$ in the collective obtained by continual drawings with replacement from the population, taking care to give each individual in the population an equal chance of appearance in the collective. It is exactly the same as with the throws of a die, when we sometimes count the appearance of a 6 as a 'success' (p), treating all other numbers as a 'failure' (q), and sometimes regard all three even numbers as a 'success', treating the odd numbers as a 'failure'. Limits of confidence can be separately calculated for the proportion estimated for every class in a sample.

There is, however, another kind of information which we have not so far discussed. Some attributes are quantitative and represent a measurement of each individual in the population. For example, people have attributes of height, weight, age, and income; business firms have attributes of total annual sales, numbers of employees, and value of capital employed; electric light bulbs have the attribute of hours of life; fabric has the attribute of a breaking strength measurable in pounds; and so on. A sample can be exceptionally useful as a method of collecting this kind of information, not only because sampling is quicker and less costly than a complete survey, but also because the accuracy of each measurement can be more easily ensured in a sample than in a population as a whole. *Notwithstanding any sampling error, an average measurement obtained by sample can be more reliable than one calculated from all items in the population.*

How do we compute the sampling error of this kind of quantitative attribute? The standard deviation described in Chapter VII provides a basis for assessing the confidence we can have in accepting the proportions given in a sample as being a true reflection of the proportions which possess the same attributes in the population. But we are not dealing now with the presence or absence of an attribute but with the average of an attribute which is part of every one of the members of a population. If these measurements are all ranked according to their magnitude, we get a frequency distribution in which the classes represent every different measurement and the frequencies are the numbers of items in the population which have each of those measurements. Each measurement can, in turn, be regarded as a class in which that particular measurement is present, all the others together being a class from which that particular measurement is absent. The total number of classes in such a frequency distribution depends on the precision with which the measurements are stated. If we classify people by their ages expressed as completed years reached on a given date, we count into a particular age class (e.g. 39 years) all those who have just reached that age on that date and all those who are less than one year older, including those who are merely one day off their 40th birthday. People who are slightly different in age would, in fact, be counted as being of the same age. On the other hand, if we take a more precise measurement of age—say, to the nearest day—we would have many times the number of classes. It is the same with other kinds of measurement. We can measure the inside diameter of a metal cylinder to the nearest one-sixteenth, one-hundredth or one-thousandth of an inch, according to the accuracy required for the purpose of the cylinder and the accuracy possible with the measuring tool. But, whatever degree of precision we take, we can always conceive of each measurement as being also a range of more precise measurements. The number of people having the height of 68 inches, measured to the nearest inch, is also the same as the number who have heights ranging from 67·5 to 68·4 inches, measured to the nearest one-tenth of an inch. And if we take measurements to the nearest tenth of an inch we can say that the class 67·8 inches is the same as the range of measurements from 67·75 to 67·84 inches, assuming the heights had been measured to the nearest one-hundredth of an inch instead. Thus, measurements are variables which in theory have the character of continuity discussed on page 46 in Chapter VI, although, in practice, they are finite with differences determined by the limitations of measuring instruments and by the desirability and higher cost of increased precision.

The limits of confidence which we now seek concern the estimates obtained from samples of the average value of a particular kind of attribute—a measurement—belonging to *all* items in a population; whereas the confidence limits previously described in this book concerned the estimates obtained from samples of the *frequency* or the *relative frequency* of an attribute possessed by some but not possessed by other members of a population. If, in order to estimate the average heights of men and women in a population of 50,000,000 persons, we take a sample of the heights of 1000 persons, we are concerned not only with the total number of ways in which 50 million persons can be combined 1000 at a time but with an equal number of estimates of the average height. The amount of arithmetic is clearly formidable. In sampling with replacement, each item in the population can be combined with any other item *including itself* so the number of ways in which the members of a population (N) can be combined in samples of size n is N^n. Therefore the number of estimates of the average height of 50 million people obtained from samples of 1000 is no less than $50,000,000^{1000}$—that is to say, 50 million multiplied by itself one thousand times! What we need to know is the shape of the frequency distribution of all these estimates of the average measurement and the probability of getting near to the true average of the whole population as we increase the size of the sample. Of course, when the sample is as large as the population it would in practice be silly not to take the population and be done with it, but we are interested in discovering a general formula, so we must keep to a random selection of items with replacement, thus creating a collective in which the probability of getting each item and, with it, each measurement, is constant. The size of sample in this collective can then be imagined to grow indefinitely.

We shall see more clearly the shape of the frequency distributions of the average measurements of all the possible samples of n drawn with replacement from a population of given size N if we plot them on charts. Unfortunately, as we have seen, the total number of samples taken from a real population is likely to be so great that the task of estimating the average measurement for every sample would be much too onerous. In order to demonstrate the role of probability in obtaining measurements by sample, we need to have a very small population of the kind we had in games of chance—such as the six sides of a die— and to work with exceptionally small samples. We will therefore imagine our population to consist merely of ten persons and we will see what happens when we try to estimate their average height (to the nearest inch) by means of a sample. Details of this imaginary population are given in Table 16.

It will be convenient to begin by examining the results of taking samples of two persons only. This is, of course, a ridiculously small sample, but it has the merit of providing a simple combination which can be set out in two-dimensional tabular form. We draw one person at random from the population, measure his height, replace him, then draw another and measure his height. All of the 100 combinations possible ($10^2 = 10 \times 10$) and the average height in each sample are shown in Table 17. Ten results come from combinations of two identical persons; and 90 come from combinations of two different persons, $(^{10}C_2 = 10 \times 9)$. As explained in the previous chapter, the process of random selection in sampling means that we deliberately give each member of a population an equal chance of being included in a sample, so the probability of drawing any one of the ten persons in our imaginary population is 0·1. By the multiplication principle, the probability of getting any one of the combinations of two persons is 0·01 (i.e. 0·1 × 0·1) so the chances of getting any of the average heights shown in Table 17 are equal. It will be noticed that some of these average measurements are the same. We can, therefore, simplify the presentation of these results by constructing a frequency distribution in which the different average measurements are ranked according to their magnitude and the number of samples having each is placed alongside. The proportion of all possible samples which have a particular average measurement represents the probability of getting that answer, for it is the addition of the probabilities of getting every one of the samples of 2 which have that average value. This frequency distribution is shown in Table 18 and in Chart 11.

Let us now increase the size of sample to 3. To do this we combine each of the ten members of the original population with each of the 100 combinations of two, making a total of 1000 possible samples of three $(10^3 = 10 \times 10 \times 10)$ all of which have equal probabilities of 0·001. It would be much too complicated to show all of these combinations separately but it is not necessary because we are really interested in the probability of getting the different possible average measurements and these are shown most clearly in a frequency distribution which summarizes the sample results. This distribution of all the samples of 3 persons is shown in Table 19 and in Chart 12.

It will be noted that the shape of this probability distribution of the average measurements obtained from samples of 3 persons looks rather like that of a Normal Curve. We must, however, examine some larger samples since we wish to ascertain the effect of sample size on the accuracy of the measurements obtained. Because the amount of computation grows very quickly with larger samples, we will raise the number in the sample by one only, to 4 persons. Even so, there are no fewer than 10,000 equally probable combinations of 4 obtainable from the population of 10 persons $(10^4 = 10 \times 10 \times 10 \times 10)$ and these produce the 37 different estimates of the average height of those ten persons shown in Table 20. In Chart 13 we see the pattern of these results. The average measurements obtained in samples of 5—a total of $10^5 = 100,000$ equally probable combinations—are given in Table 21 and shown in Chart 14.

What conclusions can we draw from these tables and charts? The reader will notice that, even with these very small samples, the graphs of the frequency distributions become narrower as the size of the sample is increased. This means that the proportion of all possible samples having average measurements which fall within a given margin of error increases as the size of sample is increased. For example, if the limits are set at $1\frac{1}{2}$ inches each side of the known true average height of 66·5 inches we find that our chances of getting answers within these limits improve in the following way:

Sample size	Probability of getting measurements from 65 to 68 inches
2	0·58
3	0·67
4	0·73
5	0·78

This can be stated another way round. There is a given probability of finding the true answer within a margin of error which, itself, diminishes in size as the sample is enlarged. Thus, if we are satisfied with a 0·95 chance of getting the true answer, the margin of error will be ±3·5 with samples of 2 (i.e. 95% of all possible samples will possess measurements between 63 and 70 inches), ±3·3 with samples of 3, ±2·75 with samples of 4, and ±2·5 with samples of 5.

We can imagine what happens as we go on taking larger and larger samples. There is no end to it. The random series created by the drawings with replacement from the population is infinite; and the true average measurement of all the items of the population is a limiting value to which the sample averages converge. As *n* tends to infinity, the margin of error tends to vanish altogether for a *given probability* of finding the true answer. This given probability may be 0·95, or 0·995 or 0·9995, but, ultimately, with a large enough sample, we shall have an error which is insignificantly small. Similarly, the probability of being within a *given margin of error* tends to certainty as *n* tends to infinity, however narrow the limits are set. Although we can never be quite sure that we shall get the correct average measurement by sample, we know that we shall get very close to it with a large enough sample.

The limiting value of the true average measurement in the collective is not itself a probability. It is the average of measurements weighted by the limiting values of the frequencies of getting every measurement in the population. These limiting values of the *frequencies* of getting particular measurements are the probabilities in the collective and these frequencies have a binomial distribution which is approximated by the Normal curve. Thus, in our imaginary population in Table 16, the probability of getting a person with the height of (say) 67 inches is 0·1 not because one-tenth of the population have this height but because this is the limiting value in the collective formed by a random selection which, with replacement, is designed to give each member an equal chance of inclusion in samples. It is clear why the average measurement in the collective tends to the true average; the measurements attributable to each member of the population are then weighted by probabilities which are equal to the proportions found in the population. What we also need to know is whether the *distribution* of errors in estimating the average measurement (i.e. the frequencies of the differences between all sample averages and the true average), like the distribution of errors in proportions obtained from samples, approaches a limit in the form of the Normal curve because, if it does, we have at hand a ready means of computing the reliability of sample results in terms of the standard deviation.

We can find out quite easily. If we convert to standard deviations the differences between the average measurement in each sample of a given size and the average measurement of all possible samples of that size (e.g. column 2 of Table 21), we can read off the corresponding values for the ordinates of the Normal curve, or the areas under it, from a standard table of the kind given in Table 12. This will give the figures for a Normal curve having the same central measurement and the same standard deviation as the frequency distribution of the average sample measurements. Since the Normal curve is the limit approached by the binomial theorem as *n* tends to infinity, it becomes a better approximation as the size of the sample is enlarged. Unfortunately a sample of 5 is the largest for which we have information and this is really very small. Nevertheless, a comparison of the frequency distribution of samples of 5 in Table 21 with a Normal curve having the same central value (66·5 inches) and standard deviation (1·287) will indicate whether the form of the two distributions appears to be the same.

Chart 15 shows that the probability distribution of average heights estimated from these samples in spite of their smallness follows fairly closely the shape of the Normal curve. For this chart, the ordinates of the Normal curve have been taken. The procedure is given in Table 22. First, the differences of each of the 46 estimates of the average height from the average of all possible samples—which, as we know, is the same as the true average height of the population—are converted to standard deviations (column (2) of Table 22). For example, the 3795 samples giving an estimate of the average height of 67·8 inches overstate the true figure by 1·3 inches, which is equivalent to an error of 1·01 standard deviations (1·3 ÷ 1·287). The abscissa (horizontal) scale of the chart is in standard deviations measured each side of the central correct value. Next, from a more detailed version of Table 14 we ascertain the ordinates of the Normal curve corresponding to the abscissa points in column (2). It does not matter whether we use the actual values of y in the formula for the Normal curve or these figures expressed as a proportion of the maximum ordinate value, because the ratios are identical and they have to be turned into relative frequencies for comparison with the ordinates of the frequency distribution of samples of 5. These relative frequencies are probabilities which together add up to unity. The 46 values of the ordinates of the Normal curve are, therefore, converted to probabilities as shown in column (4) of Table 22, and the corresponding figures for samples of 5 are placed alongside in column (5). Let us work through one row of figures. An error of 1·3 inches in estimating the average height is 1·01 in terms of standard deviations and the y value of the Normal curve ordinate at this point (1·01) is 0·23955. This is 0·037 of the total of all the 46 Normal curve ordinate values (0·23955 ÷ 6·43366), so this represents a kind of theoretical probability to be expected if the finite sampling distribution is approximated by the Normal curve. In fact, the probability of getting an error of 1·3 inches in samples of 5 persons is 0·038, so the approximation is very close.

We can, however, make the comparison in a different way. Although, in practice, a sampling distribution is finite, with probabilities which can be assigned to definite values of the average measurement, the Normal curve is continuous with no clear steps between one value and another, for it is the limit of the binomial theorem when n is infinitely large. Probabilities are, therefore, measured by the *area* under the curve lying between two ordinates, and these ordinates can be erected anywhere along the abscissa scale which is also continuous. With samples of 5 there are steps between contiguous average heights of 0·2 inches; and there are steps of 0·2 inches between contiguous differences between a sample average and the true average height. However, we can regard each of these average measurements and each difference from the true average as being, in effect, a class with a lower and an upper limit, as shown in columns (2) and (3) of Table 23. The probability of getting any one of these classes can be read from a standard Normal curve table as the proportion of the total area under the Normal curve lying between ordinates erected at these limits expressed as a proportion of the standard deviation of the sampling distribution. In other words, it is the area of a column having the width of the class interval (0·2 inches) divided by the standard deviation (1·287). The height of the column is determined by its position under the curve. The procedure is shown in Table 23. Only the upper limit of a class need be expressed in terms of the standard deviation, because the upper limit of one class is the lower limit of the next as we move from the central value, 0, either leftwards for the negative errors, or rightwards for the positive errors. This is done in column (4).

The area contained in a column under the Normal curve between the central maximum ordinate and one erected at 0·16 standard deviations away is 0·06356 of the total area under the curve. This represents an estimate of the probability of getting samples of 5 which have an average height of not more than 0·2 inches below the true figure, and there is a similar probability of getting samples which have an average of not more than 0·2 inches above the true figure. Therefore, we can say that, if the sampling distribution is Normal in shape, the approximate probability of getting a sample of 5 giving an average height of between 66·3 and 66·7 inches—66·5 plus or minus 0·2—is 0·12712 (i.e. 0·06356 × 2). This compares with a probability of 0·12 taken from the frequency distribution of all possible samples of 5.

The area between the central ordinate and one erected at a point 0·31 standard deviations away is 0·12172 and, since the area between the central ordinate and one at 0·16 standard deviations is 0·06356, the area in a column under the Normal curve between ordinates erected at 0·16 and 0·31 must be 0·12172 minus 0·06356, which equals 0·05816. This, then, is an approximation of the probability of getting samples of 5 with an error between 0·2 and 0·4 inches below the true figure, for comparison with a probability of 0·05875 computed from the frequency distribution of all possible samples of 5. For so small a sample, the approximation is very good. In columns (6) and (7) of Table 23 it can be seen that the approximations derived from the Normal curve are close to the probabilities calculated from the actual sampling distribution. Indeed, if we are interested in knowing the chance of obtaining an average height of between 64·5 and 68·5 inches in these small samples of 5 (i.e. a margin of error of not more than 2 inches each side of the true figure) we could quickly estimate it as being 0·87886 from a table of the Normal curve, whereas a detailed calculation of the relative frequencies of average measurements in all possible samples of 5 from a population of 10 would give the probability as 0·87834.

Clearly the Normal curve gives a good approximation of the distribution of sampling errors of measurements with this small sample, and it would be even better with much larger samples. Once again, the Normal curve is most useful where it is most needed.

We are, however, faced with the same dilemma that we had when dealing with the meaning of probability and the estimation of proportions by sample: we do not know the standard deviation of the average measurements in all possible samples of size n. Moreover, the formulae $\sqrt{\dfrac{pq}{n}}$ and $\sqrt{pqn}$ are of no relevance here because they relate, respectively, to the standard deviation of a proportion and the standard deviation of the number of items or events which possess or do not possess a particular attribute, whereas we are now interested in a quantitative attribute which is part of all the members of a population and part of every member in a sample.

The only standard deviation we know is the standard deviation of measurements obtained in the sample taken, and we would expect this to reflect the standard deviation of the measurements of the items in the population, subject to a sampling error. Clearly this is so because, in the random series created by repeated sampling with replacement, the limiting value of the average measurement is identical with the average measurement of the population, and the limiting value of the relative frequencies of getting each different measurement is also identical with the proportions of those measurements in the population, because each member of the population is deliberately given an equal chance of inclusion in the sample. Therefore, the larger the sample, the better the probability that the standard deviation of the sample measurements will be close to the standard deviation of the population.

But what we need, in order to make use of the Normal curve for assessing the reliability of a sample result, is the standard deviation of the *average* measurements of all possible samples of a given size; and this is not the same as the standard deviation of the individual measurements obtained in a sample (however large) or as the standard deviation of all the individual measurements in all possible samples. In fact, the standard deviation of the individual measurements in the frequency distribution of samples of a given size is the same as the standard deviation of the measurements in the population being sampled. Again, this follows from the basic assumption of equal probabilities for the individual members of the population, because this means that the proportions of each measurement in the population are reproduced exactly in the sampling distribution. A quick examination of Table 17 will confirm this, Here we have the different ways in which all ten measurements in the imaginary population of persons (Table 16) can be drawn with replacement to make samples of 2. It will be seen that there are 18 samples containing the measurement of 62 inches once combined with another measurement, and there is one sample containing this measurement twice, so it appears 20 times in a total of 200 appearances of all ten measurements in the collection of 100 possible samples of 2 from this population ($N^n \times n = 10^2 \times 2$)—i.e. one-tenth, which is the same as the proportion in the population. It can be shown that the proportion of each of the other nine heights is similarly one-tenth in the sampling distribution as in the population. Since each height in Table 17 appears 20 times, the variance is simply the squares of the deviations in column (4) of Table 16 multiplied by 20 instead of by one, making a total of 1650 (instead of 82·5) divided by a total frequency of 200, which is also 20 times larger, thus giving an identical dividend (the variance) of 8·25. Therefore, the standard deviation, which is the square root of the variance, is 2·8723 for the individual measurements in the frequency distribution of samples just as it is for the measurements in the population.

Let us recapitulate for a moment. The Normal curve approximates very well the frequency distribution of the measurements in all possible samples of a given size and the standard deviation of the *average* measurements in this distribution therefore provides an easy and satisfactory measure of the limits of confidence in estimating an average measurement from a sample. However, the only standard deviation we know is the one obtained from the measurements of the items in the sample taken and this provides a good estimate of the standard deviation of the measurements in the population being samples; but this standard deviation is *not* the same as the standard deviation of the *average* measurements in all possible samples of a given size, which is what we want, although it happens to be identical with the standard deviation of the individual measurements of the items in all these samples. So, in order to make use of the standard deviation estimated from the sample taken, we need to know whether there is any mathematical relationship between the standard deviation of the individual measurements and the standard deviation of the average measurements in the frequency distribution of all possible samples of the given size. Intuitively we would expect one, because we know that, as the value of n is increased, the first standard deviation remains unchanged (being identical with the standard deviation of the population) whereas the second standard deviation gets smaller.

There is, in fact, a relationship between the two standard deviations and, not surprisingly, it is the size of the sample (value of n) which is the connecting link. The following are the variances of the average heights obtainable in all possible samples of 2, 3, 4, and 5 persons drawn from the imaginary population of ten persons (see Tables 17, 18, 19 and 20):

Value of n	Variance of average heights in the sampling distributions
2	4·125
3	2·750
4	2·063
5	1·650

The variance of the heights in the population is 8·25, which is the same as the variance of the individual heights in all the samples in each of the four sampling distributions, and if we divide this by the value of n, we get the variance shown above, thus:

Variance of heights in population (V_p)	Value of n	$V_p \div n$
8·25	2	4·125
8·25	3	2·750
8·25	4	2·063
8·25	5	1·650

The connection may therefore be given a general form, applicable to all sizes of sample, by saying that the variance of the average measurements in the frequency distribution of samples of a given size (V_s) is equal to the variance of the population measurements (V_p), divided by the size of the sample, n, thus:

$$V_s = \frac{V_p}{n} .$$

The relationship between the two standard deviations is now obvious. A standard deviation is the square root of a variance, so

$$\sqrt{V_s} = \frac{\sqrt{V_p}}{\sqrt{n}}$$

$$= \sigma_s = \frac{\sigma_p}{\sqrt{n}}$$

which, in words, is: the standard deviation of the average measurements in a sampling distribution (σ_s) equals the standard deviation of the measurements in the population (σ_p) divided by the square root of the size of the sample ($\sqrt{n}$). In practice, we rarely know the population standard deviation and take, instead, an estimate of it from the measurements obtained in a sample.

TABLE 1
THE ARITHMETICAL TRIANGLE

Series of terms 2^0 --- 2^{24}

m (No. of terms)	Series of terms																									Total of terms
1	1																									1 2^0
2	1	1																								2 2^1
3	1	2	1																							4 2^2
4	1	3	3	1																						8 2^3
5	1	4	6	4	1																					16 2^4
6	1	5	10	10	5	1																				32 2^5
7	1	6	15	20	15	6	1																			64 2^6
8	1	7	21	35	35	21	7	1																		128 2^7
9	1	8	28	56	70	56	28	8	1																	256 2^8
10	1	9	36	84	126	126	84	36	9	1																512 2^9
11	1	10	45	120	210	252	210	120	45	10	1															1,024 2^{10}
12	1	11	55	165	330	462	462	330	165	55	11	1														2,048 2^{11}
13	1	12	66	220	495	792	924	792	495	220	66	12	1													4,096 2^{12}
14	1	13	78	286	715	1,287	1,716	1,716	1,287	715	286	78	13	1												8,192 2^{13}
15	1	14	91	364	1,001	2,002	3,003	3,432	3,003	2,002	1,001	364	91	14	1											16,384 2^{14}
16	1	15	105	455	1,365	3,003	5,005	6,435	6,435	5,005	3,003	1,365	455	105	15	1										32,768 2^{15}
17	1	16	120	560	1,820	4,368	8,008	11,440	12,870	11,440	8,008	4,368	1,820	560	120	16	1									65,536 2^{16}
18	1	17	136	680	2,380	6,188	12,376	19,448	24,310	24,310	19,448	12,376	6,188	2,380	680	136	17	1								131,072 2^{17}
19	1	18	153	816	3,060	8,568	18,564	31,824	43,758	48,620	43,758	31,824	18,564	8,568	3,060	816	153	18	1							262,144 2^{18}
20	1	19	171	969	3,876	11,628	27,132	50,388	75,582	92,378	92,378	75,582	50,388	27,132	11,628	3,876	969	171	19	1						524,288 2^{19}
21	1	20	190	1,140	4,845	15,504	38,760	77,520	125,970	167,960	184,756	167,960	125,970	77,520	38,760	15,504	4,845	1,140	190	20	1					1,048,576 2^{20}
22	1	21	210	1,330	5,985	20,349	54,264	116,280	203,490	293,930	352,716	352,716	293,930	203,490	116,280	54,264	20,349	5,985	1,330	210	21	1				2,097,152 2^{21}
23	1	22	231	1,540	7,315	26,334	74,613	170,544	319,770	490,314	646,646	705,432	646,646	490,314	319,770	170,544	74,613	26,334	7,315	1,540	231	22	1			4,194,304 2^{22}
24	1	23	253	1,771	8,855	33,649	100,947	245,157	490,314	817,190	1,144,066	1,352,078	1,352,078	1,144,066	817,190	490,314	245,157	100,947	33,649	8,855	1,771	253	23	1		8,388,608 2^{23}
25	1	24	276	2,024	10,626	42,504	134,596	346,104	735,471	1,307,504	1,961,256	2,496,144	2,704,156	2,496,144	1,961,256	1,307,504	735,471	346,104	134,596	42,504	10,626	2,024	276	24	1	16,777,216 2^{24}

H

TABLE 2

DEMONSTRATION THAT THE PROPORTION OF TOTAL SUCCESSES IN A BINOMIAL DISTRIBUTION $= p$

A Sample of 100,000 sets of 10 dice

No. of sixes (r) (1)	Probability $^{10}C_r q^{n-r} p^r$ (2)	*No. of sets of 10 in sample (3)	Total number of sixes $= (1) \times (3)$ (4)
0	0·16151	16,151	
1	0·32301	32,301	32,301
2	0·29071	29,071	58,142
3	0·15504	15,504	46,512
4	0·05427	5,427	21,708
5	0·01302	1,302	6,510
6	0·00217	217	1,302
7	0·00025	25	175
8	0·00002	2	16
9	†	†	—
10	†	†	—
	1·00000	100,000	166,666

* Assuming the binomial distribution, $(\frac{5}{6} + \frac{1}{6})^{10}$, is faithfully reproduced.

† A much larger sample than 100,000 is required to get either of these.

The total number of results $= 100,000 \times 10 = 1,000,000$. As 166,666 of these are sixes, the proportion is clearly:

$$\frac{166,666}{1,000,000} = 0 \cdot 166 = \tfrac{1}{6} = p$$

∴ $100,000(\frac{5}{6} + \frac{1}{6})^{10}$ results in 166,666 sixes and 833,334 other sides.

TABLE 3

EXPANSION OF $(q+p)^{10}$ WHERE $q = \frac{1}{2}$ AND $p = \frac{1}{2}$

No. of heads r	Proportion of heads: $\frac{r}{n}$	Probability $^nC_r p^r q^{n-r}$	No. of heads r	Proportion of heads: $\frac{r}{n}$	Probability $^nC_r p^r q^{n-r}$
0	0·0	·0010	6	0·6	·2051
1	0·1	·0098	7	0·7	·1172
2	0·2	·0439	8	0·8	·0439
3	0·3	·1172	9	0·9	·0098
4	0·4	·2051	10	1·0	·0010
5	0·5	·2460			

TABLE 4

EXPANSION OF $(q+p)^{24}$ WHERE $q = \frac{1}{2}$ AND $p = \frac{1}{2}$

No. of heads r	Proportion of heads: $\frac{r}{n}$	Probability $^nC_rp^rq^{n-r}$	No. of heads r	Proportion of heads: $\frac{r}{n}$	Probability $^nC_rp^rq^{n-r}$
< 3		·0002	13	0·542	·1488
			14	0·583	·1169
3	0·125	·0001	15	0·625	·0779
4	0·166	·0006	16	0·667	·0438-
5	0·208	·0025	17	0·708	·0206
6	0·250	·0080	18	0·750	·0080
7	0·292	·0206	19	0·792	·0025
8	0·333	·0438	20	0·833	·0006
9	0·375	·0779	21	0·875	·0001
10	0·417	·1169			
11	0·458	·1488	> 21		·0002
12	0·500	·1612			

TABLE 5

EXPANSION OF $(q+p)^{100}$ WHERE $q = \frac{1}{2}$ AND $p = \frac{1}{2}$

No. of heads r	Proportion of heads: $\frac{r}{n}$	Probability $^nC_rp^rq^{n-r}$	No. of heads r	Proportion of heads: $\frac{r}{n}$	Probability $^nC_rp^rq^{n-r}$
< 32		·0001	51	0·51	·0780
32	0·32	·0001	52	0·52	·0735
33	0·33	·0002	53	0·53	·0666
34	0·34	·0005	54	0·54	·0579
35	0·35	·0009	55	0·55	·0485
36	0·36	·0016	56	0·56	·0389
37	0·37	·0027	57	0·57	·0301
38	0·38	·0045	58	0·58	·0223
39	0·39	·0071	59	0·59	·0159
40	0·40	·0108	60	0·60	·0108
41	0·41	·0159	61	0·61	·0071
42	0·42	·0223	62	0·62	·0045
43	0·43	·0301	63	0·63	·0027
44	0·44	·0389	64	0·64	·0016
45	0·45	·0485	65	0·65	·0009
46	0·46	·0579	66	0·66	·0005
47	0·47	·0666	67	0·67	·0002
48	0·48	·0735	68	0·68	·0001
49	0·49	·0780			
50	0·50	·0796	> 68		·0001

TABLE 6

EXPANSION OF $(q+p)^{10}$ WHERE $q = \frac{5}{6}$ AND $p = \frac{1}{6}$

No. of sixes r	Proportion of sixes: $\dfrac{r}{n}$	Probability $^nC_rp^rq^{n-r}$	No. of sixes r	Proportion of sixes: $\dfrac{r}{n}$	Probability $^nC_rp^rq^{n-r}$
0		·1615	6	0·6	·0022
1	0·1	·3229	7	0·7	·0003
2	0·2	·2907	8	0·8	
3	0·3	·1551	9	0·9	
4	0·4	·0543	10	1·0	
5	0·5	·0130			

TABLE 7

EXPANSION OF $(q+p)^{24}$ WHERE $q = \frac{5}{6}$ AND $p = \frac{1}{6}$

No. of sixes r	Proportion of sixes: $\dfrac{r}{n}$	Probability $^nC_rp^rq^{n-r}$	No. of sixes r	Proportion of sixes: $\dfrac{r}{n}$	Probability $^nC_rp^rq^{n-r}$
0		·0126	8	0·333	·0237
1	0·042	·0604	9	0·375	·0084
2	0·083	·1389	10	0·417	·0025
3	0·125	·2036	11	0·458	·0006
4	0·167	·2138	12	0·500	·0001
5	0·208	·1711			
6	0·250	·1083			
7	0·292	·0557	>12		·0002

TABLE 8

EXPANSION OF $(q+p)^{100}$ WHERE $q = \frac{5}{6}$ AND $p = \frac{1}{6}$

No. of sixes r	Proportion of sixes: $\frac{r}{n}$	Probability $^nC_r p^r q^{n-r}$	No. of sixes r	Proportion of sixes: $\frac{r}{n}$	Probability $^nC_r p^r q^{n-r}$
<5		·0001	18	0·18	·0970
5	0·05	·0003	19	0·19	·0838
6	0·06	·0009	20	0·20	·0679
7	0·07	·0025	21	0·21	·0517
8	0·08	·0058	22	0·22	·0371
9	0·09	·0118	23	0·23	·0252
10	0·10	·0214	24	0·24	·0162
11	0·11	·0350	25	0·25	·0098
12	0·12	·0520	26	0·26	·0057
13	0·13	·0703	27	0·27	·0031
14	0·14	·0874	28	0·28	·0016
15	0·15	·1002	29	0·29	·0008
16	0·16	·1065	30	0·30	·0004
17	0·17	·1052	31	0·31	·0002
			>31		·0001

TABLE 9

TRANSFORMATION OF BINOMIAL DISTRIBUTION
TO NORMAL CURVE $(\frac{1}{2}+\frac{1}{2})^{10}$

Deviation from the average $(r-\bar{r})$	$(r-\bar{r}) \div \sigma$ $= \frac{(r-\bar{r})}{1\cdot58114}$	Probability of deviation $^nC_r q^{n-r} p^r$	Column (3) × σ
(1)	(2)	(3)	(4)
−5	−3·1623	·0010	·0016
−4	−2·5300	·0098	·0155
−3	−1·8974	·0439	·0694
−2	−1·2649	·1172	·1853
−1	−0·6325	·2051	·3243
0	—	·2460	·3890
+1	+0·6325	·2051	·3243
+2	+1·2649	·1172	·1853
+3	+1·8974	·0439	·0694
+4	+2·5300	·0098	·0155
+5	+3·1623	·0010	·0016

TABLE 10

TRANSFORMATION OF BINOMIAL DISTRIBUTION TO NORMAL CURVE $(\frac{1}{2}+\frac{1}{2})^{24}$

Deviation from average $(r-\bar{r})$	$(r-\bar{r})\div\sigma$ $=\dfrac{(r-\bar{r})}{2\cdot44949}$	Probability of deviation ${}^{n}C_{r}q^{n-r}p^{r}$	Column (3) × σ
(1) −9	(2)	(3)	(4)
−9	−3·67423	·0001	·000245
−8	−3·26599	·0006	·001470
−7	−2·85774	·0025	·006124
−6	−2·44949	·0080	·019596
−5	−2·04124	·0206	·050459
−4	−1·63299	·0438	·107287
−3	−1·22474	·0779	·190814
−2	−0·81650	·1169	·286333
−1	−0·40825	·1488	·364469
0	—	·1612	·394850
+1	+0·40825	·1488	·364469
+2	+0·81650	·1169	·286333
+3	+1·22474	·0779	·190814
+4	+1·63299	·0438	·107287
+5	+2·04124	·0206	·050459
+6	+2·44949	·0080	·019596
+7	+2·85774	·0025	·006124
+8	+3·26599	·0006	·001470
+9	+3·67423	·0001	·000245
+9			

TABLE 11

TRANSFORMATION OF BINOMIAL DISTRIBUTION
TO NORMAL CURVE $(\frac{1}{2}+\frac{1}{2})^{100}$

Deviation from the average $(r-\bar{r})$	$(r-\bar{r})\div\sigma$ $=\dfrac{(r-\bar{r})}{5}$	Probability of deviation $^{n}C_{r}q^{n-r}p^{r}$	Column (3) $\times$ σ
(1)	(2)	(3)	(4)
−18			
−18	−3·6	·0001	·0005
−17	−3·4	·0002	·0010
−16	−3·2	·0005	·0025
−15	−3·0	·0009	·0045
−14	−2·8	·0016	·0080
−13	−2·6	·0027	·0135
−12	−2·4	·0045	·0225
−11	−2·2	·0071	·0355
−10	−2·0	·0108	·0540
− 9	−1·8	·0159	·0795
− 8	−1·6	·0223	·1115
− 7	−1·4	·0301	·1505
− 6	−1·2	·0389	·1945
− 5	−1·0	·0485	·2425
− 4	−0·8	·0579	·2895
− 3	−0·6	·0666	·3330
− 2	−0·4	·0735	·3675
− 1	−0·2	·0780	·3900
0	—	·0796	·3980
+ 1	+0·2	·0780	·3900
+ 2	+0·4	·0735	·3675
+ 3	+0·6	·0666	·3330
+ 4	+0·8	·0579	·2895
+ 5	+1·0	·0485	·2425
+ 6	+1·2	·0389	·1945
+ 7	+1·4	·0301	·1505
+ 8	+1·6	·0223	·1115
+ 9	+1·8	·0159	·0795
+10	+2·0	·0108	·0540
+11	+2·2	·0071	·0355
+12	+2·4	·0045	·0225
+13	+2·6	·0027	·0135
+14	+2·8	·0016	·0080
+15	+3·0	·0009	·0045
+16	+3·2	·0005	·0025
+17	+3·4	·0002	·0010
+18	+3·6	·0001	·0005
+18			

TABLE 12

TABLE OF ORDINATES OF AND AREA UNDER THE
NORMAL CURVE

Distance (x) from the middle term as a proportion of the standard deviation, σ $= \dfrac{x}{\sigma}$	Ordinates		Area under the curve between the middle term and an ordinate erected at $\dfrac{x}{\sigma}$
	y	As a proportion of the maximum ordinate	
0·0	0·39894	1·00000	·00000
0·1	0·39695	·99501	·03983
0·2	0·39104	·98020	·07926
0·3	0·38139	·95600	·11791
0·4	0·36827	·92312	·15542
0·5	0·35207	·88250	·19146
0·6	0·33322	·83527	·22575
0·7	0·31225	·78270	·25804
0·8	0·28969	·72615	·28814
0·9	0·26609	·66689	·31594
1·0	0·24197	·60653	·34134
1·1	0·21785	·54607	·36433
1·2	0·19419	·48675	·38493
1·3	0·17137	·42956	·40320
1·4	0·14973	·37531	·41924
1·5	0·12952	·32531	·43319
1·6	0·11092	·27804	·44520
1·7	0·09405	·23575	·45543
1·8	0·07895	·19790	·46407
1·9	0·06562	·16448	·47128
2·0	0·05399	·13534	·47725
2·1	0·04398	·11025	·48214
2·2	0·03547	·08892	·48610
2·3	0·02833	·07100	·48928
2·4	0·02239	·05614	·49180
2·5	0·01753	·04394	·49379
2·6	0·01358	·03405	·49534
2·7	0·01042	·02612	·49653
2·8	0·00792	·01984	·49744
2·9	0·00595	·01492	·49813
3·0	0·00443	·01111	·49865

TABLE 13

CONFIDENCE LIMITS FOR THE TOSSES OF
A COIN

Size of sample n	Standard deviation, σ	The limits within which the values of $\frac{r}{n}$ are likely to be found in 99·73% of all samples: i.e. $0·50 \pm 3\sigma$	
		Lower	Upper
100	0·05000	0·35	0·65
200	0·03536	0·39392	0·60608
300	0·02887	0·41339	0·58661
400	0·02500	0·42500	0·57500
500	0·02236	0·43292	0·56708
600	0·02041	0·43877	0·56123
700	0·01890	0·44330	0·55670
800	0·01769	0·44693	0·55307
900	0·01666	0·45002	0·54998
1,000	0·01581	0·45257	0·54743
2,000	0·01118	0·46646	0·53354
3,000	0·00913	0·47261	0·52739
4,000	0·00791	0·47627	0·52373
5,000	0·00707	0·47879	0·52121
6,000	0·00645	0·48065	0·51935
7,000	0·00598	0·48206	0·51794
8,000	0·00559	0·48323	0·51677
9,000	0·00527	0·48419	0·51581
10,000	0·00500	0·48500	0·51500
20,000	0·00354	0·48938	0·51062
30,000	0·00289	0·49133	0·50867
40,000	0·00250	0·49250	0·50750
50,000	0·00223	0·49331	0·50669
60,000	0·00204	0·49388	0·50612
70,000	0·00189	0·49433	0·50567
80,000	0·00177	0·49469	0·50531
90,000	0·00167	0·49499	0·50501
100,000	0·00158	0·49526	0·50474
1,000,000	0·00050	0·49850	0·50150
100,000,000	0·00005	0·49985	0·50015
10,000,000,000	0·000005	0·499985	0·500015

TABLE 14

STANDARD DEVIATIONS FOR DIFFERENT VALUES OF
p AND n

Value of p	Standard deviation, $\sigma = \sqrt{\dfrac{pq}{n}}$			
	$n = 100$	$n = 500$	$n = 1000$	$n = 5000$
·05	·0218	·00975	·00689	·00308
·10	·0300	·01342	·00949	·00424
·15	·0357	·01597	·01129	·00505
·20	·0400	·01789	·01265	·00566
·25	·0433	·01937	·01369	·00612
·30	·0458	·02049	·01450	·00648
·35	·0477	·02133	·01508	·00675
·40	·0490	·02191	·01550	·00693
·45	·0498	·02225	·01573	·00704
·50	·0500	·02236	·01581	·00707
·55	·0498	·02225	·01573	·00704
·60	·0490	·02191	·01550	·00693
·65	·0477	·02133	·01508	·00675
·70	·0458	·02049	·01450	·00648
·75	·0433	·01937	·01369	·00612
·80	·0400	·01789	·01265	·00566
·85	·0357	·01597	·01129	·00505
·90	·0300	·01342	·00949	·00424
·95	·0218	·00975	·00689	·00308

TABLE 15

ANNUAL ESTIMATES OF POPULATION
(000's) FOR THE UNITED KINGDOM,
1963

Total population	53,673	
	Males	*Females*
Total	26,036	27,637
Age groups		
Under 5	2326	2206
5–9	2029	1926
10–14	2016	1914
15–19	2137	2059
20–24	1700	1718
25–29	1695	1639
30–34	1712	1664
35–39	1761	1752
40–44	1859	1887
45–49	1664	1733
50–54	1766	1854
55–59	1634	1769
60–64	1315	1584
65–69	963	1352
70–74	684	1086
75–79	441	784
80–84	230	459
85 and over	104	252

Sources: Registrars General

TABLE 16

IMAGINARY POPULATION OF THE HEIGHTS
OF TEN PERSONS

Height to nearest inch	Frequency f	Deviation from the average height d	Squares of the deviations in col. (3) d^2
(1)	(2)	(3)	(4)
62	1	−4·5	20·25
63	1	−3·5	12·25
64	1	−2·5	6·25
65	1	−1·5	2·25
66	1	−0·5	·25
67	1	+0·5	·25
68	1	+1·5	2·25
69	1	+2·5	6·25
70	1	+3·5	12·35
71	1	+4·5	20·25
	10		82·50

Average height in the population = 66·5 inches

Variance $\left(V = \dfrac{\Sigma d^2 f}{10} \right) = \dfrac{82·50}{10} = 8·25$

Standard deviation $(\sqrt{V} = \sigma) = 2·8723$

TABLE 17

THE AVERAGE HEIGHTS OF ALL POSSIBLE COMBINATIONS OF
TWO PERSONS FROM THE IMAGINARY POPULATION
(TABLE 16)

					Inches					
Height	62	63	64	65	66	67	68	69	70	71
62	62·0	62·5	63·0	63·5	64·0	64·5	65·0	65·5	66·0	66·5
63	62·5	63·0	63·5	64·0	64·5	65·0	65·5	66·0	66·5	67·0
64	63·0	63·5	64·0	64·5	65·0	65·5	66·0	66·5	67·0	67·5
65	63·5	64·0	64·5	65·0	65·5	66·0	66·5	67·0	67·5	68·0
66	64·0	64·5	65·0	65·5	66·0	66·5	67·0	67·5	68·0	68·5
67	64·5	65·0	65·5	66·0	66·5	67·0	67·5	68·0	68·5	69·0
68	65·0	65·5	66·0	66·5	67·0	67·5	68·0	68·5	69·0	69·5
69	65·5	66·0	66·5	67·0	67·5	68·0	68·5	69·0	69·5	70·0
70	66·0	66·5	67·0	67·5	68·0	68·5	69·0	69·5	70·0	70·5
71	66·5	67·0	67·5	68·0	68·5	69·0	69·5	70·0	70·5	71·0

TABLE 18

FREQUENCY DISTRIBUTION OF SAMPLES OF 2

Average height in inches	Deviation of average height from the true average	Number of samples	Probability
(1)	(2)	(3)	(4)
62·0	−4·5	1	·01
62·5	−4·0	2	·02
63·0	−3·5	3	·03
63·5	−3·0	4	·04
64·0	−2·5	5	·05
64·5	−2·0	6	·06
65·0	−1·5	7	·07
65·5	−1·0	8	·08
66·0	−0·5	9	·09
66·5	—	10	·10
67·0	+0·5	9	·09
67·5	+1·0	8	·08
68·0	+1·5	7	·07
68·5	+2·0	6	·06
69·0	+2·5	5	·05
69·5	+3·0	4	·04
70·0	+3·5	3	·03
70·5	+4·0	2	·02
71·0	+4·5	1	·01
		100	1·00

Average height in all 100 samples = 66·5 inches

Variance $\left(V = \dfrac{\Sigma d^2 f}{100} \right)$ = 4·125

Standard deviation ($\sqrt{V} = \sigma$) = 2·031

TABLE 19

FREQUENCY DISTRIBUTION OF SAMPLES OF 3

Average height in inches	Deviation of average height from the true average (d)	Number of samples (f)	Probability
(1)	(2)	(3)	(4)
62·0	−4·50	1	·001
62·3	−4·16̇	3	·003
62·6	−3·83̇	6	·006
63·0	−3·50	10	·010
63·3	−3·16̇	15	·015
63·6	−2·83̇	21	·021
64·0	−2·50	28	·028
64·3	−2·16̇	36	·036
64·6	−1·83̇	45	·045
65·0	−1·50	55	·055
65·3	−1·16̇	63	·063
65·6	−0·83̇	69	·069
66·0	−0·50	73	·073
66·3	−0·16̇	75	·075
66·6	+0·16̇	75	·075
67·0	+0·50	73	·073
67·3	+0·83̇	69	·069
67·6	+1·16̇	63	·063
68·0	+1·50	55	·055
68·3	+1·83̇	45	·045
68·6	+2·16̇	36	·036
69·0	+2·50	28	·028
69·3	+2·83̇	21	·021
69·6	+3·16̇	15	·015
70·0	+3·50	10	·010
70·3	+3·83̇	6	·006
70·6	+4·16̇	3	·003
71·0	+4·50	1	·001
		1000	1·000

Average height in all 1000 samples = 66·5 inches

Variance $\left(V = \frac{\Sigma d^2 f}{1000} \right)$ = 2·58358

Standard deviation ($\sqrt{V} = \sigma$) = 1·607

TABLE 20

FREQUENCY DISTRIBUTION OF SAMPLES OF 4

Average height in inches	Deviation of average height from the true average (d)	Number of samples (f)	Probability
(1)	(2)	(3)	(4)
62·00	−4·50	1	·0001
62·25	−4·25	4	·0004
62·50	−4·00	10	·0010
62·75	−3·75	20	·0020
63·00	−3·50	35	·0035
63·25	−3·25	56	·0056
63·50	−3·00	84	·0084
63·75	−2·75	120	·0120
64·00	−2·50	165	·0165
64·25	−2·25	220	·0220
64·50	−2·00	282	·0282
64·75	−1·75	348	·0348
65·00	−1·50	415	·0415
65·25	−1·25	480	·0480
65·50	−1·00	540	·0540
65·75	−0·75	592	·0592
66·00	−0·50	633	·0633
66·25	−0·25	660	·0660
66·50	—	670	·0670
66·75	+0·25	660	·0660
67·00	+0·50	633	·0633
67·25	+0·75	592	·0592
67·50	+1·00	540	·0540
67·75	+1·25	480	·0480
68·00	+1·50	415	·0415
68·25	+1·75	348	·0348
68·50	+2·00	282	·0282
68·75	+2·25	220	·0220
69·00	+2·80	165	·0165
69·25	+2·75	120	·0120
69·50	+3·00	84	·0084
69·75	+3·25	56	·0056
70·00	+3·50	35	·0035
70·25	+3·75	20	·0020
70·50	+4·00	10	·0010
70·75	+4·25	4	·0004
71·00	+4·50	1	·0001
		10,000	1·0000

Average height in all 10,000 samples = 66·5 inches

Variance $\left(V = \dfrac{\Sigma d^2 f}{10,000} \right)$ = 2·0625

Standard deviation ($\sqrt{V} = \sigma$)   = 1·436

TABLE 21
FREQUENCY DISTRIBUTION OF SAMPLES OF 5

Average height in inches	Deviation of average height from the true average (d)	Number of samples (f)	Probability
(1)	(2)	(3)	(4)
62·0	−4·5	1	·00001
62·2	−4·3	5	·00005
62·4	−4·1	15	·00015
62·6	−3·9	35	·00035
62·8	−3·7	70	·00070
63·0	−3·5	126	·00126
63·2	−3·3	210	·00210
63·4	−3·1	330	·00330
63·6	−2·9	495	·00495
63·8	−2·7	715	·00715
64·0	−2·5	996	·00996
64·2	−2·3	1340	·01340
64·4	−2·1	1745	·01745
64·6	−1·9	2205	·02205
64·8	−1·7	2710	·02710
65·0	−1·5	3246	·03246
65·2	−1·3	3795	·03795
65·4	−1·1	4335	·04335
65·6	−0·9	4840	·04840
65·8	−0·7	5280	·05280
66·0	−0·5	5631	·05631
66·2	−0·3	5875	·05875
66·4	−0·1	6000	·06000
66·6	+0·1	6000	·06000
66·8	+0·3	5875	·05875
67·0	+0·5	5631	·05631
67·2	+0·7	5280	·05280
67·4	+0·9	4840	·04840
67·6	+1·1	4335	·04335
67·8	+1·3	3795	·03795
68·0	+1·5	3246	·03246
68·2	+1·7	2710	·02710
68·4	+1·9	2205	·02205
68·6	+2·1	1745	·01745
68·8	+2·3	1340	·01340
69·0	+2·5	996	·00996
69·2	+2·7	715	·00715
69·4	+2·9	495	·00495
69·6	+3·1	330	·00330
69·8	+3·3	210	·00210
70·0	+3·5	126	·00126
70·2	+3·7	70	·00070
70·4	+3·9	35	·00035
70·6	+4·1	15	·00015
70·8	+4·3	5	·00005
71·0	+4·5	1	·00001
		100,000	1·00000

Average height in all 100,000 samples = 66·5 inches

Variance $\left(V = \dfrac{\Sigma d^2 f}{100,000} \right)$ = 1·656936

Standard deviation $(\sqrt{V} = \overline{\sigma})$ = 1·287

TABLE 22

Difference of sample average from true average d	Difference expressed as standard deviations d	Normal curve ordinate at point shown in column (2)	Theoretical probability of sample average	Probability with samples of 5 shown in Table 21
(1)	(2)	(3)	(4)	(5)
−4·5	3·50	·00087	·00014	·00001
−4·3	3·34	·00151	·00024	·00005
−4·1	3·19	·00246	·00038	·00015
−3·9	3·03	·00405	·00063	·00035
−3·7	2·87	·00649	·00101	·00070
−3·5	2·72	·00987	·00153	·00126
−3·3	2·56	·01506	·00234	·00210
−3·1	2·41	·02186	·00340	·00330
−2·9	2·25	·03174	·00493	·00495
−2·7	2·10	·04398	·00684	·00715
−2·5	1·94	·06077	·00945	·00996
−2·3	1·79	·08038	·01249	·01340
−2·1	1·63	·10567	·01642	·01745
−1·9	1·48	·13344	·02074	·02205
−1·7	1·32	·16694	·02595	·02710
−1·5	1·17	·20121	·03127	·03246
−1·3	1·01	·23955	·03723	·03795
−1·1	·85	·27798	·04321	·04335
−0·9	·70	·31225	·04853	·04840
−0·7	·54	·34482	·05360	·05280
−0·5	·39	·36973	·05747	·05631
−0·3	·23	·38853	·06039	·05875
−0·1	·08	·39767	·06181	·06000
+0·1	·08	·39767	·06181	·06000
+0·3	·23	·38853	·06039	·05875
+0·5	·39	·36973	·05747	·05631
+0·7	·54	·34482	·05360	·05280
+0·9	·70	·31225	·04853	·04840
+1·1	·85	·27798	·04321	·04335
+1·3	1·01	·23955	·03723	·03795
+1·5	1·17	·20121	·03127	·03246
+1·7	1·32	·16694	·02595	·02710
+1·9	1·48	·13344	·02074	·02205
+2·1	1·63	·10567	·01642	·01745
+2·3	1·79	·08038	·01249	·01340
+2·5	1·94	·06077	·00945	·00996
+2·7	2·10	·04398	·00684	·00715
+2·9	2·25	·03174	·00493	·00495
+3·1	2·41	·02186	·00340	·00330
+3·3	2·56	·01506	·00234	·00210
+3·5	2·72	·00987	·00153	·00126
+3·7	2·87	·00649	·00101	·00070
+3·9	3·03	·00405	·00063	·00035
+4·1	3·19	·00246	·00038	·00015
+4·3	3·34	·00151	·00024	·00005
+4·5	3·50	·00087	·00014	·00001
			1·00000	1·00000

I

TABLE 23

Difference of sample average from true average d	Class limits		Upper limits expressed as standard deviations y	Area between middle ordinate and class limit	Theoretical probability of sample average	Probability with samples of 5 shown in Table 21
	Lower x	Upper y				
(1)	(2)	(3)	(4)	(5)	(6)	(7)
−4·5	4·4	4·6	3·57	·49982	·00013	·00001
−4·3	4·2	4·4	3·42	·49969	·00025	·00005
−4·1	4·0	4·2	3·26	·49944	·00038	·00015
−3·9	3·8	4·0	3·11	·49906	·00065	·00035
−3·7	3·6	3·8	2·95	·49841	·00097	·00070
−3·5	3·4	3·6	2·80	·49744	·00159	·00126
−3·3	3·2	3·4	2·64	·49585	·00224	·00210
−3·1	3·0	3·2	2·49	·49361	·00351	·00330
−2·9	2·8	3·0	2·33	·49010	·00473	·00495
−2·7	2·6	2·8	2·18	·48537	·00706	·00715
−2·5	2·4	2·6	2·02	·47831	·00975	·00996
−2·3	2·2	2·4	1·86	·46856	·01219	·01340
−2·1	2·0	2·2	1·71	·45637	·01694	·01745
−1·9	1·8	2·0	1·55	·43943	·02019	·02205
−1·7	1·6	1·8	1·40	·41924	·02673	·02710
−1·5	1·4	1·6	1·24	·39251	·03037	·03246
−1·3	1·2	1·4	1·09	·36214	·03833	·03795
−1·1	1·0	1·2	0·93	·32381	·04151	·04335
−0·9	·8	1·0	0·78	·28230	·04993	·04840
−0·7	·6	·8	0·62	·23237	·05513	·05280
−0·5	·4	·6	0·46	·17724	·05552	·05631
−0·3	·2	·4	0·31	·12172	·05816	·05875
−0·1	—	·2	0·16	·06356	·06356	·06000
+0·1	—	·2	0·16	·06356	·06356	·06000
+0·3	·2	·4	0·31	·12172	·05816	·05875
+0·5	·4	·6	0·46	·17724	·05552	·05631
+0·7	·6	·8	0·62	·23237	·05513	·05280
+0·9	·8	1·0	0·78	·28230	·04993	·04840
+1·1	1·0	1·2	0·93	·32381	·04151	·04335
+1·3	1·2	1·4	1·09	·36214	·03833	·03795
+1·5	1·4	1·6	1·24	·39251	·03037	·03246
+1·7	1·6	1·8	1·40	·41924	·02673	·02710
+1·9	1·8	2·0	1·55	·43943	·02019	·02205
+2·1	2·0	2·2	1·71	·45637	·01694	·01745
+2·3	2·2	2·4	1·86	·46856	·01219	·01340
+2·5	2·4	2·6	2·02	·47831	·00975	·00996
+2·7	2·6	2·8	2·18	·48537	·00706	·00715
+2·9	2·8	3·0	2·33	·49010	·00473	·00495
+3·1	3·0	3·2	2·49	·49361	·00351	·00330
+3·3	3·2	3·4	2·64	·49585	·00224	·00210
+3·5	3·4	3·6	2·80	·49744	·00159	·00126
+3·7	3·6	3·8	2·95	·49841	·00097	·00070
+3·9	3·8	4·0	3·11	·49906	·00065	·00035
+4·1	4·0	4·2	3·26	·49944	·00038	·00015
+4·3	4·2	4·4	3·42	·49969	·00025	·00005
+4·5	4·4	4·6	3·57	·49982	·00013	·00001
					·99964	1·00000

CHART 1

The formation of permutations

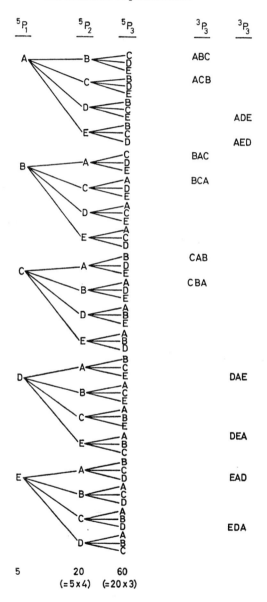

Total	5	20	60
		(= 5×4)	(= 20×3)

I*

CHART 2

Expansion of $(\frac{1}{2} + \frac{1}{2})^n$: Numbers of Heads

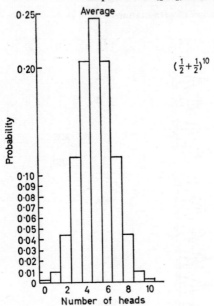

$(\frac{1}{2} + \frac{1}{2})^{10}$

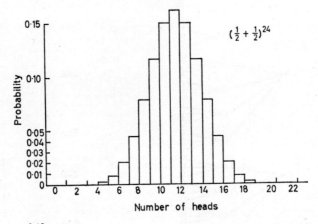

$(\frac{1}{2} + \frac{1}{2})^{24}$

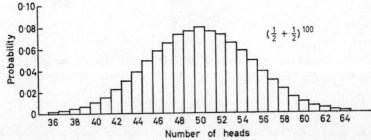

$(\frac{1}{2} + \frac{1}{2})^{100}$

CHART 3

Expansion of $(\frac{1}{2}+\frac{1}{2})^n$: Proportions of Heads

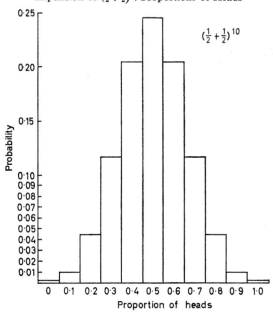

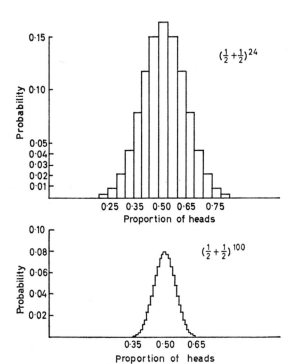

CHART 4

Expansion of $(\frac{5}{6} + \frac{1}{6})^n$: Numbers of Sixes

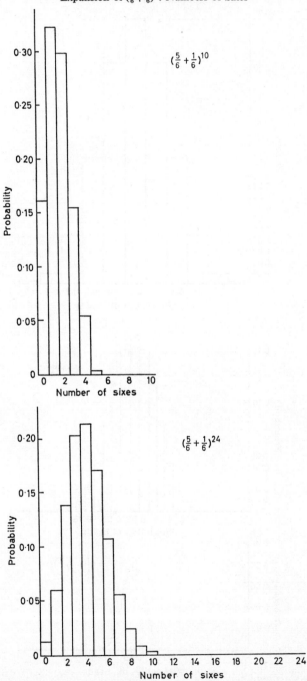

CHART 4 cont.

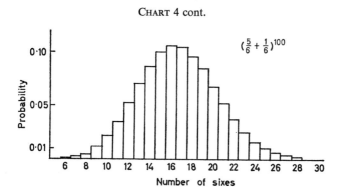

Number of sixes

CHART 5

Expansion of $(\frac{5}{6} + \frac{1}{6})^n$: Proportions of Sixes

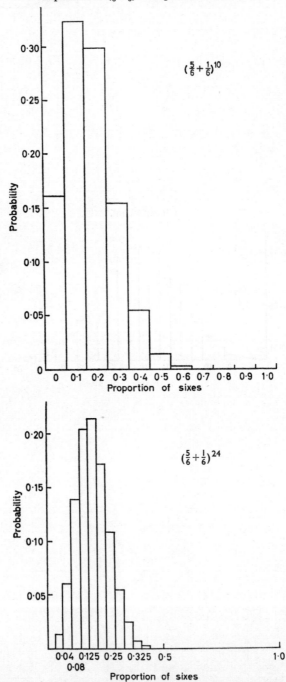

CHART 5 cont.

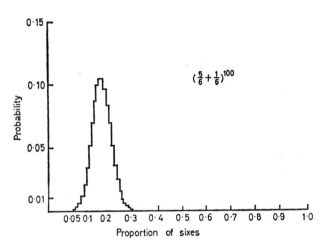

$(\frac{5}{6} + \frac{1}{6})^{100}$

CHART 6
The Birth of the Normal Curve

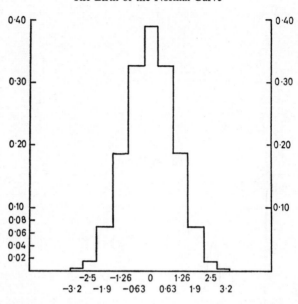

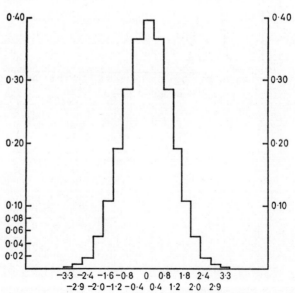

CHART 6 cont.

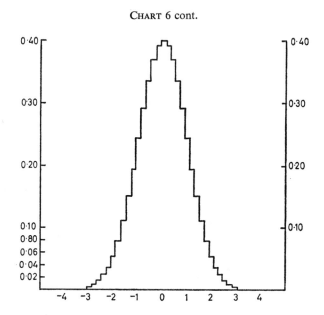

CHART 7

The Normal Curve

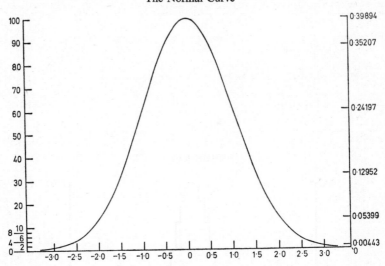

CHART 8

3 Standard Deviation Band

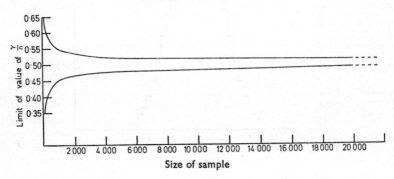

Size of sample

CHART 9

Value of S.D. as value of *p* gets larger or smaller than 0·50

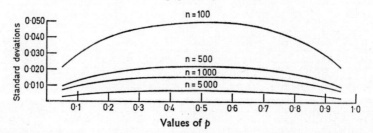

Values of *p*

CHART 10
Ratios of S.D.s

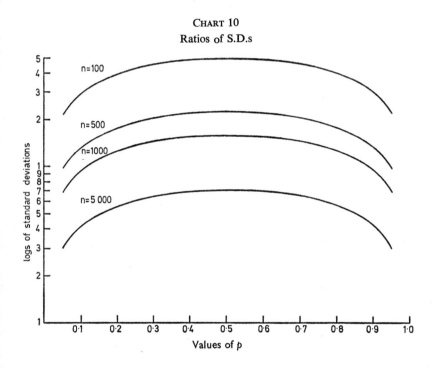

logs of standard deviations

Values of *p*

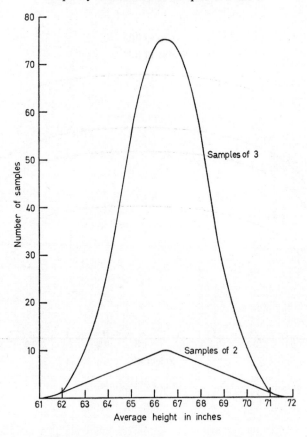

CHARTS 11 and 12
Frequency Distribution of Samples of 2 and 3

CHART 13
Frequency Distribution of Samples of 4

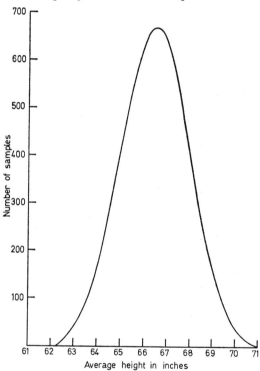

CHART 14
Frequency Distribution of Samples of 5

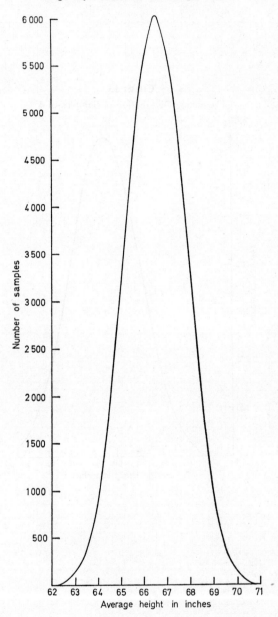

CHART 15

Probability of obtaining average heights with samples of 5 compared with the probability of obtaining the same average heights with a Normal Distribution

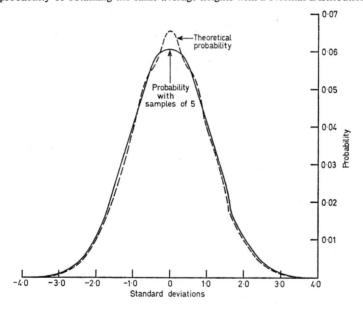

A USEFUL BIBLIOGRAPHY

1. *How to take a Chance*
   DARNELL HUTT (Penguin, 1965)

2. *Probability and Statistics for Everyman*
   IRVING ADLER (Dobson Books Ltd., 1963)

3. *Games, Gods and Gambling*
   F. N. DAVID (Charles Griffin, 1962)

4. *Probability and Scientific Inference*
   G. SPENCER-BROWN (Longmans, Green & Co., 1957)

5. *Principles of Statistical Techniques*
   P. G. MOORE (Cambridge University Press, 1958)

6. *An Introduction to the Theory of Statistics*
   YULE and KENDALL (Charles Griffin, 1958)

7. *Sampling Techniques*
   WILLIAM G. COCHRAN (Chapman & Hall Ltd., 1963)

8. *Chance and Choice*
   L. HOGBEN (Max Parrish, 1950)

INDEX